Inventing *Modern*

Inventing
Modern

Growing up with X-rays, Skyscrapers, and Tailfins

JOHN H. LIENHARD

OXFORD
UNIVERSITY PRESS

2003

OXFORD
UNIVERSITY PRESS

Oxford   New York
Auckland   Bangkok   Buenos Aires   Cape Town   Chennai
Dar es Salaam   Delhi   Hong Kong   Istanbul   Karachi   Kolkata
Kuala Lumpur   Madrid   Melbourne   Mexico City   Mumbai   Nairobi
São Paulo   Shanghai   Taipei   Tokyo   Toronto

Published by Oxford University Press, Inc.
198 Madison Avenue, New York, New York 10016
www.oup.com

Oxford is a registered trademark of Oxford University Press

Library of Congress Cataloging-in-Publication Data
Lienhard, John H., 1930–
Inventing modern : growing up with x-rays, skyscrapers, and tailfins
/by John H. Lienhard.
p. cm.   Includes bibliographical references and index.
ISBN 0-19-516032-0
1. United States—Civilization—20th century. 2. Technological
innovations—Social aspects—United States—History—20th century.
3. Technology—Social aspects—United States—History—20th century.
4. Science—Social aspects—United States—History—20th century.
5. Material culture—Social aspects—United States—History—20th century.
6. Lienhard, John H., 1930– —Childhood and youth. I. Title.
E169.1 .L53945 2003   303.48'3'097309045—dc21   2002156634   Rev.

9 8 7 6 5 4 3 2 1
Printed in the United States of America
on acid-free paper

Contents

Preface

I use the title "Inventing *Modern*" as a reminder of the creativity that brought a radically new epoch into being, a century ago. I take the liberty of making a proper noun of the word *Modern* since I am inventing my own take on that epoch. Historians and literary scholars have given the adjective "modern" various dimensions of specificity, and I have no wish to intrude upon any of these canonical meanings. My *Modern* (the noun) is that strange state into which I was born. It is a distinctly American *Modern*. It is a highly personal *Modern*. And it is a condition that dissolved during my early adult life.

My intention here is neither historical nor literary in the formal sense of either word. It is, instead, an attempt to understand a now-bygone culture by seeing it as the product of new technology. *Modern* is a condition that echoes from my childhood. Born in 1930, I grew up in a world transformed. Just how it had been transformed could be understood at the time only in terms of the broadest outward symptoms. Here was a world dramatically different from any that had ever been, and a single word floated above it all. That word was "modern." Everyone used it. Everyone knew what it meant. No one knew what it meant.

The old word "modern" had already taken on wholly new implications just before I was born. Now it surrounded us. For example, to tell us that they had replaced their old outhouses with indoor plumbing, motor courts put the code word "modern" on their signs. *Modern* was our new state of being.

To understand any epoch we must inevitably reclaim some sense of its technological and artistic texture. Many historians have done that very well, but none can ever do it completely. Even the *Modern* that we all saw at the time was an ephemeral state, one dancing with new X-rays, radium, and radio waves—with Art Deco, Bauhaus, and skyscrapers.

We saw *Modern* in Burma-Shave signs, the now-abandoned Teepee Motel down in Wharton County, Texas, and Grauman's Chinese Theater. *Modern* was Fritz Lang's movie *Metropolis* and the talking pictures that followed it. My *Modern* was the Chrysler Airflow, Buck Rogers, and the DC-3. I last saw my *Modern* speeding off into the mists of the fifties on great automobile tailfins.

Science certainly provides one red thread, perhaps the key thread, running through all the more evident facets of *Modern*. When relativity theory and quantum mechanics departed from everything we had believed to constitute reality, they told us that a time had arrived when nothing would be the same—that we faced change of a magnitude hitherto unimaginable.

Yet the immensity of the transformation that occurred just after 1900 is actually cloaked by the accomplishments that brought it about. To see it, we have to open ourselves to the kitsch and commonplace material world in which we struggled to express the mystery of the transformation. We were a people buoyed by our new ability and strength. Not until we had twice watched all our new technology feeding the monster of war did the worm of doubt finally eat into the heart of that childlike assurance.

But the dark side of *Modern* is another story for some postmodern day. For the moment, let us go back to look for the spirit of a world being turned into something that no one had imagined possible. I shall attempt to interpret my primal awareness of that world just as it took form—the *Modern* that infused my childhood and had been the driving imperative of my parents' early adult lives as well.

I carve my zoetropic representation of *Modern* around an arbitrary selection of events. A thousand specific leaps of human energy would write much the same story; it would be foolish to attempt an encyclopedic accounting. *Modern* was the sum of far greater social, cultural, and artistic change after 1900 than anyone could ever recount in a single book. I select certain glimpses and try to put them under the lens of new technologies and new science.

Finishing a book takes self-isolating focus, and it lures you into thinking that you've been working alone. Only as you emerge from that cocoon of labor to smell fresh ink upon the final stack of pages do you realize how many people have stood with you and how greatly they've shared in the work.

My wife, Carol Ann, has read the entire book several times and, in those instances when she has not had specific comments, she has allowed me to read the infinitely useful expression on her face. Three University of Houston colleagues have read the entire manuscript and made cogent commentary: Dr. John E. Fadell, M.D. Anderson Library; Dr.

Robert Zaretsky, University of Houston, Department of Modern and Classical Languages; and Margaret Culbertson, head of the Art and Architecture Library. Ms. Culbertson also provided a great deal of *a priori* content advice and provided a number of the illustrations.

My thanks to University of Houston colleagues, Dr. Dorothy Z. Baker, English Department, and Dr. Steven Mintz, History Department, for very helpful critical readings of selected chapters. I also owe much to several anonymous reviewers who critically reviewed early versions of the first chapters.

Then there is a web of subtler contribution. How many times did I draw my history colleagues, Dr. Sarah Fishman-Boyd and Dr. Richard Blackett, into lunchtime conversations and left with ideas clarified? The title of this book came out of an *ad hoc* committee that I gathered to formulate the gift CD for contributors to Houston public radio station KUHF-FM, in the year 2000. That was the first draft of this book. Beyond the CD, many ideas in this book first appeared on my nationally distributed KUHF program, *The Engines of Our Ingenuity.* Capella Tucker, producer of the program, general manager John Proffitt, and the entire KUHF staff, have been most supportive. University colleagues and administrators in engineering, history, the library, and many other colleges, have been constantly helpful.

Finally, my gratitude to production editor Helen Mules and all the other good people at Oxford University Press. My special thanks to Executive Editor Kirk Jensen. He has been an enormously positive force—the iron fist in the velvet glove, always driving me to do better.

2003   John H. Lienhard

Inventing *Modern*

1

1846: Great-Grandpa and Manifest Destiny

We finished the twentieth century living in what literary people were first to identify as a postmodern age. Seventy years earlier, I had been born into a world that we all knew was unambiguously modern. The word "modern" was so ubiquitous that I am still caught off guard when people use it to describe a time gone by. How did that world get away from me!

My father had good cause to use the word. Born in 1893 and raised in the small Swiss/Austrian/German farming community of Nauvoo, Illinois, he told me stories of the emergent twentieth century. He told about the day he heard an unearthly noise coming from the road as he walked through a cornfield. He ran to the road and intercepted a strange carriage, moving without the aid of horses. He talked about returning from school another day to find his mother shouting into what looked like a plain wooden box, mounted on the hallway wall.

While the automobile and the telephone blindsided him, he had some small warning about airplanes. Vague rumors of the Wright brothers' accomplishment had reached bucolic Nauvoo before my father actually saw his first airplane. Yet, only a scant 15 years after Kitty Hawk, he was to be found in France flying Nieuports, Sopwith Camels, and SPADs.

Modern reached us with stunning suddenness after 1900. Indeed much of what has been written about the modern world is necessarily touched with, at least, a patina of autobiography. Marshall Berman began his fine analysis of this cultural epoch, *All That Is Solid Melts Into Air,* in 1982, with the words "This is far from a confessional book. Still . . ."[1] It is precisely that pause, and then that word *still,* which gives his analysis its narrative power.

Any of us who are old in the early years of the twenty-first century were raised just as that epoch crested. It had been a tsunami of change that began in the latter nineteenth century. A visitor from Europe in the High Middle Ages, suddenly transported into the nineteenth-century American West, would have recognized most of the technology around him—windmills, waterwheels, the western saddle, hard liquor, and death by hanging, to name just a few elements.[2]

Now that near-medieval life was about to turn into something seemingly incommensurate with all that had preceded it. The technologies that transformed us after 1900 had new physical principles woven through them. We had finally brought electricity, process chemistry, and internal combustion under control, and now we would leap forward.

Still, the tidal wave of *Modern* had gathered itself for a long time. Mary Shelley was one of the visionaries who, long before 1900, warned that change was coming. And she warned us that change does not come out of nothing. In the introduction to the 1819 edition of her novel *Frankenstein: Or the Modern Prometheus*, she wrote, "Everything must have a beginning. That beginning must be linked to something that went before. The Hindoos give the world an elephant to support it, but the elephant stands on a tortoise. Invention [is not created] out of the void, but out of chaos. The materials must first be there."[3]

Neither can we simply enter the front door of *Modern*; we too need to look at the chaos from which it took form. Since we need a starting point, I shall arbitrarily choose 1846 as a watershed year and pick up my own explicitly confessional narrative there.[4]

That was the year when America turned from a struggling new nation into the new bully on the block. It was the year when America claimed a "Manifest Destiny" to own the continent. We had annexed the sovereign nation of Texas in 1845. Since we wanted present-day California, Arizona, New Mexico, Nevada, and Utah, we went to war with Mexico in April of 1846. We lost 2,000 men in action, 12,000 more to disease, and we acquired all that land.

One participant in that campaign was my great-grandfather, (Johann) Heinrich Lienhard, whom I use here as a serviceable everyman in my sketch of premodern America. He was a minor player, but a major chronicler, of the westward movement. He traveled through the unfolding drama and left behind a rich record of the texture of those times.

Johann Heinrich Lienhard, 1822–1903. (Lienhard family photo)

Heinrich Lienhard reconstructed his epic diary later in life. Each of the many portions of his journal have now been published in critical historical editions.[5] The original handwritten document covers his childhood in Switzerland, his journey to America, and his extensive travels in the American frontier between 1843 and 1850.

Heinrich emigrated from Switzerland to New Orleans in 1843. He was only 21 years old and he was looking for adventure. On April 21, 1846, Heinrich left St. Louis and set off for California, where another Swiss, Captain John Augustus Sutter, had founded the community that we call Sutter's Fort. Almost immediately after Heinrich reached California that October, he joined the U.S. Army, which was trying to secure its presence down the California coast. He needed the enlistment bonus to repay a companion for the costs of outfitting their trek.

The Mormons were simultaneously being driven out of the Illinois town that would later become Heinrich's home. Their main group began the trek to the West just a month after Heinrich did, but they had to winter over in Nebraska. They didn't reach the Great Salt Lake until the following year. Once there, they claimed a vast area around and including present-day Utah as a U.S. territory.

The adventurers who went into the West regarded its expanses as theirs for the taking. Texas colonizers had created a new nation under the Lone Star flag. The Mormons briefly called their territory *Deseret* and took a beehive as their emblem. Even California had, amid all the chaos, claimed to be a sovereign nation. California took the image of a bear as its symbol. Before gold was discovered on Sutter's property, he had called his holdings *New Helvetia*, or New Switzerland.

Heinrich had been raised to be a farmer, and he too set out to find his piece of freedom and independence within this fertile wilderness. He began his journey to California with a steamboat ride up the Missouri River from St. Louis to Independence. Independence was as far west of the Mississippi as steamboats had yet penetrated. Beyond it, he had to go on foot. And so our attention shifts to one of the primary new technologies of the western expansion. Let us look a bit more closely at the steamboat.

Robert Fulton had patented his steamboat 37 years earlier, and steamboats had been the first form of powered transportation to begin connecting the vast American continent. By 1846, steamboats, railways, and canals had all begun commerce in materiel to and from America's burgeoning new manufacturing centers. Rail was still in its infancy, and canals were impractical in the vast reaches of the American West. But steamboats had just recently evolved into a practical form.

The first western steamboats had been built in Brownsville, Pennsylvania, on the Monongahela River just south of Pittsburgh, by a Fulton-based monopoly in 1811. The group meant to create river traffic along the Monongahela/Ohio/Mississippi system.

Artist's conception of Fulton's *Clermont*, from a 1919 children's book.[6] (Note the relatively deep hull and the back-up presence of a sail.)

While textbooks often go too far in crediting Fulton with inventing the steamboat, they neglect to mention that he was first to build what would evolve into the all-important Mississippi riverboat. Fulton, however, built steamboats the way people built ships, with deep hulls and keels.

The price of being first is usually trouble, and he had his share. His boats turned huge profits for a few trips. Then they broke down and suffered accidents. Fulton was able to send his boats downriver to New Orleans. But on the way back they had only enough power to reach Natchez, where the river still ran broad and quiet. Fulton's early boats were also more vulnerable to snags and sandbars.

Fulton began steam-driven traffic on the Mississippi, then died only four years later. Today, the name of Henry Shreve overshadows him along the western river system. As captain of a deep-hulled keelboat, Shreve managed to make the first New Orleans-to-Louisville trip in 1815. Then, the next year, he turned to building steamboats. He made his first boat for the Brownsville group. And he really opened up the river trade when one of his new boats managed to make the first run from New Orleans all the way upriver to Pittsburgh.[7]

Fulton had used the older Watt engines in his boats. Shreve, on the other hand, began tinkering with the weight distribution in boats and with engine selection. He turned to the now-familiar flat-bottomed hull design. To accommodate a steam engine in a shallow-draft boat, he made a particularly daring move. He adopted the smaller, high-pressure engine that had recently been developed by American millwright Oliver Evans. By mid-century his steamboats had finally become the fully evolved riverboat we see today in replicas made for river-town visitors.

Henry Shreve built his name by surpassing Fulton in the steamboat-building business. However, the honor of being namesake to the city of Shreveport did not come directly from that accomplishment. That city lies far up the Red River, not the Mississippi, and it was cut off from the ocean by an ancient logjam, 140 miles in length. Long after he had become a major riverboat builder, it was Shreve who finally managed to clear that strange natural obstruction. When he did, the Red River finally gave northwestern Louisiana, western Arkansas, northern Texas, and southern Oklahoma access to the ocean.

Heinrich Lienhard had his first exposure to steamboats in 1843, when he debarked from the sailing vessel *Narragansett* at the mouth of the Mississippi delta. From that point, a deep-hulled, steam-powered tugboat, the *Black Star,* took him and his fellow emigrants on a two-day journey to New Orleans. There he boarded a small wood-burning riverboat, the *Meridian,* for the journey to St. Louis. And, in St. Louis, he found a lively, bustling town that had already made a major industry of steamboat building.

Between 1876 and 1884, Mark Twain wrote three great epics about riverboats and the Mississippi: *Tom Sawyer,* much of *Life on the Mississippi,* and, finally, *Huckleberry Finn.* He wrote that river into our imaginations. An old travel book entitled *America Illustrated* and published in 1883 (just before *Huckleberry Finn*) particularly catches my attention. It offers a contemporary view of that Mississippi and emphasizes how powerfully the river was imprinted upon the American consciousness by the 1880s.[8]

A long section in *America Illustrated* traces the Mississippi, not from its official headwaters in Lake Itasca, but from a pool further north that fed Itasca. It then follows the river down to St. Anthony Falls, just above the Minneapolis and St. Paul river ports. Riverboats had access from that natural obstruction all the way south to New Orleans.

That was Mark Twain's Mississippi, and its rich texture was centered on Missouri. However, Twain was caught in a time warp. He confused the river of the 1840s with that of the 1880s. Huck Finn and Tom Sawyer were children of the 1840s, yet their stories are filled with anachronisms simply because the world had changed, and Mark Twain (far more the futurist than the historian) had failed to register all the changes. By 1874, the great third-of-a-mile-long St. Louis Bridge had linked Missouri to

The St. Louis Bridge, as shown in *America Illustrated,* 1883.

Illinois, and Twain's bucolic world was giving way to our more familiar Midwestern America.

We never quite see large-scale change taking place. While that new bridge was in place, it said no more about revolution than does a new computer on my desk today. People soon forget having had to wait half a day for the ferry to cross the Mississippi. Within a year or two, back in the 1980s, I'd forgotten what it meant to work with a 300-kb hard drive, or even that I had ever owned such a stone-age machine. Can I really be said to *remember* my childhood neighborhood, where no one was aware of shower stalls, domestic air conditioning, or electric refrigerators? It takes an act of concentration to summon up that world again. Change, even rapid and radical change, is oddly invisible; it really does come to us on little cat feet.

Mark Twain's flat-bottomed riverboats were not yet fully evolved when either Heinrich or Huck Finn traveled the river. The Mississippi of the 1840s was more primitive than Mark Twain romanticized it to be in the late nineteenth century. Another great novelist, Charles Dickens, reached the river a year before Heinrich did, and his account is shocking. Little romance can be found in Dickens's description of the town of Cairo, Illinois, where the Ohio enters the Mississippi, He wrote:

> At length, upon the morning of the third day, we arrived at a spot so much more desolate than any we had yet beheld, that the forlornest places we had passed, were, in comparison with it, full of interest. At the junction of the two rivers, on ground so flat and low and marshy, that at certain seasons of the year it is inundated to the house-tops, lies a breeding-place of fever, ague, and death; vaunted in England as a mine of Golden Hope, and speculated in, on the faith of monstrous representations, to many people's ruin. A dismal swamp, on which the half-built houses rot away: cleared here and there for the space of a few yards; and teeming, then, with rank unwholesome vegetation, in whose baleful shade the wretched wanderers who are tempted hither, droop, and die, and lay their bones; the hateful Mississippi circling and eddying before it, and turning off upon its southern course a slimy monster hideous to behold; a hotbed of disease, an ugly sepulchre, a grave uncheered by any gleam of promise: a place without one single quality, in earth or air or water, to commend it: such is this dismal Cairo.[9]

When Heinrich Lienhard reached Cairo, he also wrote about it: "We landed on the right bank of the mouth of the Ohio, near an unusable old steamboat. Only a few huts were still visible in the darkness. Adjoining these lay much swamp and mud, and this was Cairo, a place which was later destined to achieve great importance." That description had, of course, been tempered by hindsight. But his and Dickens's descriptions match. Did Heinrich use hindsight to suggest possibility in that wretched place, where Dickens did not? One might think so. However,

The Heber C. Kimball/Heinrich Lienhard home as Heinrich had modified it by the latter nineteenth century. (Photo from Lienhard family files)

only 13 years after he first saw Cairo, Heinrich returned to the central Mississippi region and bought the house in Nauvoo, Illinois, which had once belonged to Mormon leader Heber C. Kimball.[10]

That same house was eventually used as the set for a movie version of *Tom Sawyer*. The house in the movie appeared as Heinrich had arranged it by the 1870s. That was roughly the same time period in which Mark Twain had begun writing about Tom Sawyer and Huck Finn. By then, the Nauvoo house (and life within it) matched the house in Twain's story better than it reflected what Tom and Huck's daily lives would have been in the 1840s.

By the time Mark Twain wrote his major novels about the Mississippi, America had established itself, not only along the river, but in lands west of the Mississippi as well. But a generation earlier, in 1846, America's new inventive genius had just begun coming to light. The time when Tom and Huck supposedly lived was the period when the country had only begun finding its role as a major international force to be reckoned with.

In 1846, Elias Howe patented his first sewing machine—an extraordinarily complex machine made to do what most had thought only human fingers would ever be capable of doing. By 1846, America already led the world in manufacturing rifles whose parts were interchangeable.

Steam, among all the nineteenth-century transformational agents, was primal. Like riverboats, railways had yet to go beyond the Mississippi basin, but they were coming. In Heinrich Lienhard's awareness, steam was only a backdrop, not an obvious transforming presence. But it was not only powering new engines; it was also driving the human intellect to increasingly subtle uses of energy.

One of the more spectacular examples was given to us in New Orleans by Norbert Rillieux.[11] In 1846, Rillieux received a patent for a fundamental distillation process. He had applied for that patent in 1843, the same year Heinrich debarked from the *Black Star* in New Orleans. Details of Rillieux's life are sketchy, but we do know that he was born in 1806 to Constance Vivant, a freed slave, and Vincent Rillieux, a wealthy engineer. New Orleans was then roughly equal parts white citizens, free black citizens, and slaves. Common law arrangements between black and white were then commonplace, and Norbert was born into freedom.

Vincent Rillieux recognized a rare intelligence in Norbert and sent him to study engineering at the École Centrale in Paris. Norbert even stayed on as an instructor and published papers on steam power before he came back to New Orleans. But Louisiana's treatment of nonwhites had deteriorated in the years between Rillieux's birth and Heinrich's arrival.

After landing in New Orleans, Heinrich and a companion walked the streets in awe of a strange new world, and Heinrich wrote this:

> We also entered a market hall where all kinds of vegetables, fruits, fish, meat, etc. were being sold. . . . The market hall was being swept and washed by a number of people, but what kind of people were these? A mulatto woman was cleaning and had an iron band around her neck with three six-to-eight-inch long spikes bent outward. A Negress had an iron band around one of her hands which was connected to another band on her ankle by a chain. A Negro had one of his ankles held to a heavy block or piece of iron by a welded chain. Others appeared to be fettered in various ways, and it was under these conditions that they had to do their work.

That was the world to which Rillieux returned from France one year after Heinrich passed through it. While Rillieux had been in Paris he'd begun working on a problem from the Louisiana sugar plantations. The last step in making white sugar is to evaporate the water used in the refining process. That exacts a terrible cost in fuel. He used his thermodynamic knowledge to invent the *multistage evaporator*. By evaporating and condensing at successively lower pressures, he was able to use heat over and over. It was a brilliant idea.

Rillieux, caught between the pernicious forces of rising racism in America and technological conservatism in France, weighed the alternatives and went back to New Orleans to work on a prototype evaporator. It was the right decision at the right time. He went on to become a highly regarded process engineer. He prospered while his machine revolutionized sugar refining. But as the institution of slavery strengthened in the South during the years before the Civil War and the racial situation steadily worsened, Rillieux decided to go back to France.

There he ran into the still-quite-plausible assumption that America must be a technological outback. French engineers, misusing his process, made it look inadequate. They hurt the good name he had enjoyed as an engineer in America. Rillieux finally walked away from process engineering. He spent his last days in Paris studying archaeology and Egyptology while multistage evaporation went through successive mutations. Today the process is used for everything from desalting seawater to recycling water in space stations.

In the summer of 1843, when Heinrich left New Orleans and continued his odyssey up the Mississippi, he remarked that only four towns of any significance lay on the way to St. Louis. They were Baton Rouge, Natchez, Vicksburg, and Memphis. The rest were only collections of shacks. We have already seen that the river trip north was fairly easy as far as Natchez. That fact contributed to Natchez's major position in the burgeoning cotton trade. As America moved toward its inevitable civil war, Natchez displayed the greatest concentration of wealth in the country. Its mansions, many of which survived the war, are still spectacular today.

Natchez offered the Union Army little resistance. The South would make its stand against Grant's army in the terrible, losing siege of Vicksburg, 60 miles farther north. Twenty years after Heinrich journeyed past those towns, one set of players in the contest would be an important variant on the riverboat. These were the flat-bottomed steamboats, which still revealed their kinship with Heinrich's *Meridian* but were now clad in thick steel plate and heavily armed. Their vulnerable paddle wheels were out of sight, protected within the armor.

The story of the *Monitor*'s fighting to a draw with the

Artist's conception of the Union *Monitor* next to a typical fully evolved flat-bottomed steamboat, from *The Story of a Ship*.[12]

Artist's impression of a river gunboat battle during the Civil War.
(Currier and Ives print)

Merrimac/Virginia in Chesapeake Bay is now famous. But the real harbinger of modern naval warfare was the large fleet of ironclads that served an ongoing tactical role along the Mississippi and its tributaries. Shreve's riverboats had been formative agents as the conflict developed. They had helped to make commerce in cotton (and the slavery upon which that commerce was based) immensely profitable. Now their ironclad first cousins were helping to conclude that conflict.

Of course, while all this was going on, America's intellectual center still lay along the Atlantic. It would stay there for some time, all the while looking across the water to Europe. The West nevertheless had a special role as an agent of change just because it lay farther from the Old World. The very rawness of the West empowered it—freed it from the old canons of doing and making.

It was in the western reaches of America, and in people like Shreve and Rillieux, that we can isolate a new dynamic of technological change in America. Rillieux borrowed what he needed from Europe and used it to shape a radically new technological enterprise in a still-undeveloped land. The result was inevitably more than just borrowing. It was *recreating.*

So America borrowed and transmuted and did it all with stunning speed. Indeed, the new motive power of steam reached San Francisco Bay only months after Heinrich got there. In January 1847, during the Mexican-American War, Captain John Montgomery had planted an American flag on a California peninsula and called it San Francisco. Nine months later, a most remarkable little bark showed up in the bay. She was 15 days out of Sitka, Alaska, where a Russian-American company had built her. She was only 37 feet long.[13]

A borrowed locomotive engine set this boat apart from any other boat. She was the first steamboat on the California coast. Her five-man crew consisted of a captain, a Russian engineer, a cook, a deckhand, and a fireman. In San Francisco they also hired an Indian pilot who had come down from Sutter's Fort.

San Franciscans named the boat *Sitka*. For a while, the *Sitka* took passengers on excursions. Then, in November, she set out for Sacramento to "astonish the natives." The trip, intended to take only one day, actually took almost seven. Weeks later she still had not returned to San Francisco, so newspapers began speculating. One wag claimed that she had sailed into the California mountains to rescue a lost party.

That was gallows humor—a reference to the famous Donner party disaster. Heinrich had met and passed the ill-fated Donners in Wyoming during his trek in 1846. He reached the crest of the Sierras on October 1, 27 days ahead of the Donners, and made it safely across. When the Donners failed to show up in Sacramento, rescue parties set out to look for them. Meanwhile, snowbound in the Sierras, members of the group either died of starvation or survived by cannibalizing their dead. Their story has become a terrible American legend.

Of course, by the autumn of 1847, Donner survivors were living in San Francisco and reading what the local papers had to say about the *Sitka*. The *Sitka* was luckier than they—at least for a while. She finally did get back to San Francisco, only to be swamped by a storm. She sat in the shallows of the bay, with her smokestack sticking out of the water. And now one paper opined: "Should she be resuscitated . . . we sincerely hope that none of our citizens will trust themselves with a passage beyond the 'flat' she now rests upon." A year later, she was refitted as a sailboat and renamed the *Rainbow*. Her old engine found service grinding coffee.

The *Sitka/Rainbow* sank for the last time in February 1848, just after one of Sutter's people found gold near Sacramento. Now a great rush of people began finding their way to California any way they could. In autumn, the steam packet *California* set out from New York, around Cape Horn, to expand commerce with the suddenly rich Bay area.

Heinrich's memoirs make no mention of the *Sitka*, even though he was, by then, working for Sutter and knew some of its passengers. But the *Sitka* was one more invisible harbinger of vast change. He had no way of understanding what that embryonic little steamboat meant, nor could any of the pioneers have realized the extent to which change was upon them.

Two years after the *Sitka* arrived in San Francisco Bay, Sutter sent Heinrich back to Switzerland to pack up the Sutter family and move them to California. Now Heinrich had only to book passage on another steam packet, the *Panama*, which ran between San Francisco and Panama. He crossed the Isthmus of Panama on land and picked up a second packet to America's East Coast. Change had been that rapid.

On the surface of it, western pioneers were merely struggling to create the technologies of physical survival and then some level of physical amenities. They were in fact exerting technological control of their world with blinding speed. And now a new dimension of change would begin to telegraph its way from East to West.

Americans on the East Coast were asking how they might take their own place among European intellectuals in 1846, when an odd door opened up. Somewhat earlier, English aristocrat James Smithson had inexplicably left a half-million dollars to support the "increased diffusion of knowledge" in America. His intentions were not at all clear, and Congress sat on the money for 17 years. Finally, at the urging of people like John Quincy Adams, Congress took Smithson's money and created the Smithsonian Institution.

In 1996 the Smithsonian celebrated its 150th birthday with a book entitled *1846*.[14] It makes a big point of one particular event that year. the arrival of the European naturalist Louis Agassiz. With him, Agassiz brought a new scientific luster. If Americans felt they had a manifest destiny geographically and politically, they intended to have one scientifically as well. They hoped Agassiz would show the way.

Agassiz immediately met the famous electrical pioneer Joseph Henry, one of the few American scientists as well known as he. (Today, the unit of electrical inductance is called the *henry*.) Henry, the new head of the Smithsonian, set forth his plans for the institution on the last day of 1846. Its purpose would be to conduct research, publish papers, and educate Congress. It was to be the national think tank, not a museum.

Like Heinrich, Agassiz had been born in Switzerland, but he came from the other side of the mountain, both literally and figuratively. Heinrich was born on a farm outside of Zürich in 1822. He was raised in the stern Zwingli Protestant church. Agassiz was born in 1807. His father was a pastor in a fairly isolated, and intensely Protestant, enclave within a Catholic, French-speaking part of Switzerland.[15] A certain religious intensity marked both lives. In Heinrich's case, that came out in the form of moralistic dissections of much of the behavior around him. Its role in Agassiz' work was more complex.

Agassiz' brilliance was apparent at an early age, and his father sent him to Paris, where he studied with the great naturalist Georges Cuvier. When he was 22, Agassiz wrote a treatise on Brazilian fish. By the time he died, in 1873, he had become Harvard's most famous, best-loved professor and America's leading naturalist. He had supported women's equality and he had redesigned science education.

But Agassiz had also failed to see the coming collapse of many traditional ideas. He continued, for example, to fight the concept of evolution by natural selection on traditional religious grounds long after it had become an accepted part of late-nineteenth-century science.

At first, Agassiz had left a post at Switzerland's University of Neuchâtel and come to America only for a series of lectures. Then he took a post at Harvard. He soon became acquainted with a Philadelphia physician named Samuel Morton. Morton had collected a huge set of human skulls from all races and measured their brain capacity.[16] But he had juggled his data so that whites appeared to have by far the largest brains. Malays and American Indians were next, followed by the Chinese. Last of all, and far behind, were Africans.

In 1850, Agassiz went to a scientific conference in Charleston. There he used Morton's results to put forth his theory that the races had come from separate creations—that humankind is several species, not just one. Next, he hired a photographer and had slaves brought in, stripped naked, and posed.[17] Those chilling photos show work-battered and abused bodies and faces staring implacably at the lens. One man's back looks like a plowed field, crisscrossed with the scars of whippings. Agassiz saw only a lesser species. He eventually cast his lot with the North and the antislavery cause; however, he'd provided significant support for the cause of slavery.

Heinrich meanwhile lived much closer to the hard earth than Agassiz, and he received far different lessons on the nature of the human race. He writes of waiting on an island in the St. Croix River between Minnesota and Wisconsin for a lost comrade. He had just about despaired for the fellow's life when a canoe full of fearsome Chippewa warriors arrived. His first reaction was fear that he, himself, was now under threat. Then he discovered that the Chippewas had simply located the fellow and were returning him safely. Throughout his journeys you see his fear and abhorrence undergoing this kind of conversion into respect and understanding.

Somewhat later, wandering lost and hungry in the wilderness around the Wisconsin/Illinois border, Heinrich stumbled into a remote cabin occupied by a black family. Obviously surprised, he tells how they welcomed him and fed him. His Calvinistic ethic was particularly impressed by the fact that the homestead was *clean,* in a world where cleanliness lay beyond most people's expectations. Agassiz and Heinrich were being educated in very different ways, one in the East and the other in the West.

Some years later, a young William James went with Agassiz on a field trip to Brazil. In writing about Agassiz, he said he had profited greatly from the association. Not, he continues, "so much by what he says, for never did a man utter a greater amount of humbug, but by learning the way of feeling of such a vast practical engine as he is. . . . I delight to be with him."[18]

Agassiz carried his humbug to the grave, yet he enjoyed enormous respect to the very end. Shortly before he died, he made one last voyage to the Galapagos Islands, hoping to refute what Darwin had learned there about evolution. Darwin himself wrote to Agassiz' son before the ship sailed, asking him to convey his own good wishes for the trip. Agassiz had, after all, shaped the minds of people like William James. When he died, poet James Russell Lowell reflected the view of an admiring America with these lines:

> Three tiny words grew lurid as I read,
> And reeled commingling: Agassiz is dead!

So Americans labored toward what they felt was a manifest scientific destiny. America's day would come, and it would be built upon the spirit (if not on all the scientific work) of people like Agassiz.

The opening of the Smithsonian may seem like a small stitch in the vast tapestry of 1846, but it was one more way in which America's pretensions created an outcome. It represented a huge stride toward putting the country on an international stage. America, imperialist and racist, was claiming its intellectual as well as its territorial place. We had a great deal of growing up yet to do, but this could well be the year we left our infancy.

After 1846, we hurtled toward *Modern*. If missteps lay ahead, we were nevertheless on our way. And a special gravity keeps drawing me to the American West as I look for seeds of the world we have come to call modern. An odd temporal seam divides our thinking about the West. On one side of that seam is a Wild West of cowboys and Indians; on the other side is the twentieth-century West.

It is as though nothing separates those two pictures. In the 1870s most of the West was still out of reach of telegraph and railroads, and America had only the most elementary waterwheels and windmills to serve its energy needs. Heinrich was by then settled into a bucolic rural life. He grew roses, managed his farmland; he even did a turn as mayor of Nauvoo.

By 1900, that pastoral world was about to vanish. Then its disappearance was so rapid that we have to squint our eyes to find the seam. Take the Colorado town formed in 1878 and given the name *Columbia*. Postal authorities objected that this name would be confused with an existing town in California. So, in 1881, its citizens renamed it *Telluride*, after the tellurium in certain gold ores.[19] (The particular ores above Telluride actually contained no tellurium at all, but the name remains.)

By 1888 the California gold rush was over, and in Telluride, miners had settled into the energy-intensive work of tunnel mining and milling low-grade ore to get at the gold. The work was powered by small steam engines. First people burned all the local wood; then they used burros to cart in coal. Coal was costing an outrageous $40 per ton.

The Gold King Company was on the verge of going bankrupt when someone pointed out what was going on in the East. Six years earlier, Edison had put the first public electric power system on-line. He had set up the Pearl Street Station in New York City to supply direct current for his new electric lights. George Westinghouse had quickly followed suit with the first alternating-current (AC) system in Massachusetts.

The company wasted no time. In three years it had a Westinghouse-style hydroelectric power system in place. A six-foot Pelton water turbine, off in the mountains, drove two one-hundred-horsepower dynamos. Three-thousand-volt power traveled 2.5 miles, in bare copper wire, to an electric motor at the mine. Stories are told about workers waving their hats through six-foot electric arcs around the motor.

The system was installed in an environment of blizzards, avalanches, −40° cold snaps, and enormous water-flow variations—all far from any sort of technical support. In 1891 the company opened a school to train a type of rough-hewn electrical engineer to deal with the system.

It took a scant 20 years for those two small dynamos (along with the hair-raising technology of their use) to expand into the Telluride Power Company. By the early 1900s, the company was supplying 40,000 horsepower to three of the four states that form the remote four-corners area.

I caught just a whiff of the speed of the change in that world, the day I rode the Durango Silverton narrow-gauge railway to the top of the next mountain ridge—the one just east of Telluride. The amenities of twenty-first-century life vanish as you enter the mountain forest; steep, still, and forbidding.

That railway was finished in 1882, a scant nine years before the Telluride power plant was installed. Ride it and watch as highways, electricity, and hotels all vanish. Only the wild tangle of underbrush remains. That was what faced the people who went into the Rockies to dig gold—people lacking the most rudimentary timidity. Those, of course, were the same people who could bring about a seemingly instantaneous absorption of new technologies.

In 1910, only 19 years after the Telluride power plant was built, the government began what was, by far, the largest hydroelectric project America had yet seen. It was the Keokuk Dam across the Mississippi, just ten miles south of Nauvoo. Heinrich had died during the brief interim between the building of Telluride and Keokuk; and my father, earning money for college that summer of 1910, worked as a laborer on the Keokuk

The Durango Silverton Railway.
(Photo by John Lienhard)

cofferdam. The Telluride power plant had been the work of nineteenth-century pioneers. The Keokuk dam was *Modern* laying claim to the Mississippi. After it was done, any steamboats that reached Nauvoo would do so through a complex system of locks.

An even more dramatic example of that kind of leapfrogging from the primitive West into modern America may be found in the person of a woman born of Austrian-Jewish immigrants, just after the Civil War, in Placerville, California. Her name was Elizabeth Fleischmann, and Placerville was another gold rush town, about 40 miles west of Sutter's Fort, in the foothills of the Sierra Mountains.[20] Fleischmann's parents moved to San Francisco, where the family fell on hard times while she was in high school. Elizabeth had to drop out and go to work as a bookkeeper.

When Elizabeth was 28, the German physicist Wilhelm Conrad Roentgen learned about X-rays. His discovery riveted the public's imagination as no scientific discovery ever had. Do-it-yourself articles on X-ray devices reached San Francisco within a year. Elizabeth Fleischmann reacted with vision and audacity. She gave up bookkeeping and somehow managed to set up a complete X-ray system in an office on Sutter Street the year after Roentgen's discovery.

By 1900, Spanish-American War casualties were reaching San Francisco from the Philippines. They carried bullets and shrapnel in their bodies. Fleischmann was, by then, the best radiologist on the West Coast—maybe the best in the world.

She turned her Roentgen rays on the wounded with an uncanny spatial sense. She had a fine ability to triangulate on bullets lodged in the lungs or skull and to steer surgeons to them. She was able to regulate exposures to match tissue densities.

The chief of army medicine in San Francisco sent a set of her X-rays to the War Department. It was the best X-ray work the surgeon general had seen, so he went out to meet this E. Fleischmann fellow who had made them. He was astonished when he found a shy young Jewish woman. He caught his breath and went on to say that he had never seen better radiographic work.

People had no way of knowing that they had to shield themselves from X-rays in Fleischmann's time. No one knew how vicious X-rays were. Many early radiologists seriously burned their patients. Fleischmann was smart enough to avoid doing that, but X-ray-induced cancer was not so immediately obvious. It was normal practice for operators to hold their own hand in the rays to check exposures, and Fleischmann was now working long hours, year in and year out. By January 1905 that terribly overexposed arm had become cancerous and had to be amputated. She died seven months later. By then she had, ironically, started the experimental treatment of certain cancers with X-rays. Such treatment, of course, goes on today.

Much of what we know about Fleischmann comes from two *San Francisco Chronicle* articles, a 1900 profile of her work, and a 1905 obituary. After that, history closed over this pioneer woman who had emerged from the goldfields to hone the new field of radiology. It is a powerful example of the way the tidal wave of *Modern* washed over us too quickly to be seen before it had already engulfed us.

Fleischmann's story was a storm warning because it reflects something entirely new afoot. X-rays had nothing to do with steam or with steel. For that reason she helps me to see just why no one saw that we were about to be blown away by ideas unrelated to the past. Invention destroys all hope of predicting any technological future, and we can see that process once again in the great exposition of 1893.[21]

That same year Heinrich's grandson, my father, was born, and two years later X-rays were discovered. The Chicago Columbian Exposition was America's fourth major world's fair (depending upon which ones you count). America was young, strong, at peace, and feeling its oats. It was time to honor the four-hundredth anniversary of Columbus's voyage.

England had begun the cycle of world's fairs with the 1851 Crystal Palace Exhibition. In 1876 America celebrated its first centennial with the Philadelphia Fair. The fair drew the largest gate up to that time, eight million people. The 1889 Paris Exhibition, with the Eiffel Tower as its centerpiece, attracted 32 million. World fairs had become a very important business enterprise.

Now it was America's turn to create the "fair to end all fairs." Congress backed the idea, and our cities vied for the honor. In 1890, Congress anointed Chicago as the city with the best rail access, the most space, and the will to do the job. The organizers went to landscape architect Frederick Law Olmsted, designer of New York's Central Park.

Olmsted picked a swampy six-hundred-acre tract along Lake Michigan and set out to create a system of canals and lagoons. The fairgrounds were to be a Midwestern Venice. The budget was the better part of $1 billion in today's dollars. Add to that the exhibit money spent by 65,000 private, national, and international groups, and the 1893 fair was a vast enterprise. Architects, sculptors, and steelworkers converged on it.

The size of the original Ferris wheel has yet to be matched. It towered over the fair from a field that lay midway between the actual fairgrounds and nearby Washington Park.[22] Ever since then, the term *midway* has described the place where the rides are located in a fairground.

Yet the old photos offer no hint of the coming twentieth century's texture. The fair's architecture still looked like that of imperial Europe.

The original Ferris wheel, from *The Columbian Exposition Album*, 1893.

The electrical hall was filled with telegraphs and telephones, electric railways, elevators, and lighting. All of that was in place, but the fair's contributors failed utterly to anticipate radio, or the way small portable electric motors would soon emerge to change the American home and workplace.

The Transportation Building displayed bicycles, railways, and steamships, but none of the embryonic automobiles. The most popular exhibit was a display of farm windmills. The fair summarized our condition in 1893. It did not attempt to predict the future. The only real glimpse of the twentieth century seems to have been unintentional. The Women's Building was designed by 21-year-old MIT graduate Sophia Hayden. By 1893, MIT had been accepting women students for only 20 years, and it was ahead of most schools in doing so. Hayden was one of a mere handful of women architects who had qualified for the national competition for the design of the building.

The Women's Building was intended to showcase the women's clubs of America. Instead, it was filled with women's accomplishments in science, health care, literature, invention, and art. Even the building itself

The Women's Building, from *The Columbian Exposition Album*, 1893.

underscored the point. It quite unexpectedly emerged as the very heart of the now-unstoppable suffrage movement.

The fair drew 21 million people to Chicago, which then lay far to the west of most of the American population. By displaying the diverse forces that had made the country, it drew us together and helped shape a national identity. It reminded us of what we were at our best. No wonder poet Katherine Lee Bates visited the fair and returned home to write "America the Beautiful." Most startling was her phrase "Thine alabaster cities gleam." Those wonderful, evocative words described the shining pavilions of the fair—not any existing American city.

The "alabaster city," from *The Columbian Exposition Album*, 1893.

Heinrich Lienhard lived another ten years after the exposition. He died on December 19, 1903, just two days after the Wright brothers made their first flight at Kitty Hawk, North Carolina. But even that event traces to Heinrich's own backyard, for Milton Wright had sown the seeds of Kitty Hawk in Cedar Rapids, Iowa, a scant hundred miles north of Nauvoo.

Milton Wright was a bishop in the Church of the United Brethren in Christ, and during the 1880s he returned from one of his many trips spent attending to church business. This day he brought back with him a toy "helicopter." It consisted of two counter-rotating propellers, mounted on either end of a long stick, and powered by a rubber band that ran the length of the stick. It was very simple. You could make one yourself, and you'd find that it really does fly. Milton's two young sons, Orville and Wilbur, were captivated with flight from that moment.[23]

Whether Heinrich ever knew of Milton Wright, I do not know. He certainly did not learn that controllable powered flight had been achieved

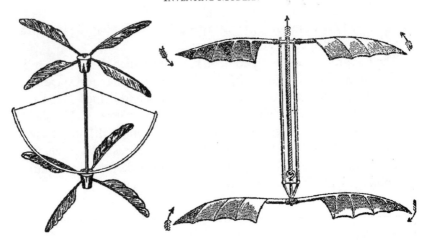

Two examples of nineteenth-century toy helicopter models, from the 1911 *Encyclopaedia Britannica*. George Cayley invented the one on the left; Alphonse Penaud, the one on the right.

during his lifetime. *Modern* had arrived, and he finished his rural life having seen only the many ambiguous early warnings that it was upon us.

America had been preparing for the twentieth century—bracing for it, in fact—for over 50 years. Yet it finally arrived too rapidly for anyone to see it coming. People had been surrounded by unprecedented change and failed to see how the character of change itself was being altered by its very rapidity. Thus, before we move into the twentieth century, we need to look more closely at the intrinsic nature of the failure to anticipate *Modern*.

2

Short-Lived Technologies: Searching for Direction

L ate-twentieth-century mathematics finally confirmed something we had always known in our bones—that no human future can ever be predicted. Insofar as we can cast the prediction of even modestly complex futures in mathematical terms, the outcome is hopelessly dependent upon specifying the present with perfect accuracy. And that, it turns out, is always impossible to do.

Think, for example, about pouring cream into tea—how the cream quickly spreads and diffuses in increasingly complicated patterns. Suppose we wanted to reproduce one such pattern. We might find a large glass tank, fill it with tea (or just water), and wait until the tea seems absolutely still. Then we might clamp a pipette filled with cream above the tank. Next, we might release one-tenth of a cubic centimeter of cream from a height of three centimeters and photograph the resulting pattern as soon as one second after impact. When that's done, we replace the tea and repeat the experiment as precisely as we can. Finally, we compare the two photos.

We find that the two patterns of the cream diffusing into tea are quite different. Even if we could write the terribly complex, time-dependent equations for the fluid motion, miniscule changes in our descriptions of the two drops hitting the surface could cause the two calculations of the diffusion to start diverging almost immediately. The size of the two drops will always differ slightly (even if only on the microscopic level). The speed of impact of the two drops hitting the water can never be precisely the same. And each droplet will suffer a slightly different distortion as it falls. Those differences can all be made tiny, but the resulting effect on the two patterns will not be tiny at all. If we wait five or ten seconds instead of one, we will see almost no relation between the two resulting patterns.

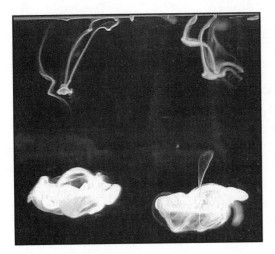

Two drops of cream entering water,
after about one second.
(Photo by John Lienhard)

This is a place where mathematics reveals and explains something we've long suspected to be true of the physical world around us. We've created a popular shorthand for that fact: we call it the *Butterfly Effect*. It is quite literally true that the movement of a butterfly's wings in Omaha, in March, *will* alter August hurricane patterns in the central Atlantic.[1] So too in human affairs; if you choose to eat cereal instead of an egg at breakfast, you literally alter the course of history.

Thus, even if we understand that we live in a time of rapid change, we can never know where change is taking us or how to use change well. People certainly saw the outward signs of rapid change during the latter half of the nineteenth century. Indeed, many felt that it was their personal manifest destiny to be among those who *drove* change. But the aggregate effect of all that change could not possibly have been evident.

Change therefore had a peculiar willy-nilly quality in the years before the twentieth century. Technological change did not gain its express-train directionality until after we had a new physics and such radical new technologies as internal combustion, a chemistry from which we could form new industrial processes at will, X-rays, and radio. Perhaps it finally took the new force of fully harnessed electricity to endow the twentieth century with its astonishing vitality.

Two kinds of technologies repeatedly emerged and then died out in the late nineteenth century. Technologies created by a linear extrapolation of an older technology soon to be replaced with something more radical were one kind. The other was technology that appeared before the needed infrastructure was there to support it. Of those technologies, many came back to claim acceptance after the infrastructure was in place.

Those two scenarios are certainly still with us today: however, we find truly dramatic examples during the mid- to late-nineteenth century— examples of short-lived technologies that briefly riveted public attention and then evaporated. Mixed in with the *keepers*—the railways, steamboats, balloon-frame houses, and telegraph—were technologies that emerged with great fanfare before vanishing. Those technologies make wonderful illustrations of the mischief caused by the Butterfly Effect.

First, two quick illustrations of the first scenario: Both are linear extrapolations, not of old *technologies*, but of old *means*. Both were created by people who had no way of seeing that a radical replacement was upon them. They were the Pony Express and cattle drives.

The Pony Express provides an excellent illustration of the brief-flame intensity of much late-nineteenth-century invention. It went into service in April 1860 and served until October of the following year. Riders covered over 1800 miles from St. Joseph to Sacramento in ten days. It took only 18 months for that flash-in-the-pan, action-based, derring-do organization to be overtaken by the electrical telegraph. The Pony Express was already taking its place as an American icon when it was blindsided by telegraphy.

The highly organized cross-country cattle drive was the most obvious means for bringing animals to slaughterhouses. It became another American icon, now woven into our literature and our legends. But no such rudimentary means *could* have lasted long. After ten years, railroads and cattle cars arrived to serve the newly settled West, and cattle drives were reduced to covering the distance to the regional rail stop.

Compare those with two examples of technologies that went beyond their supporting infrastructure: the clipper ship and Beach's secret subway.

Clipper ships were less a specific design than they were an attitude toward shipbuilding, and they lasted only a decade. Clippers ranged in size from a few hundred tons to over four thousand tons. Between one hundred and four hundred were built, depending on which design we choose to give the name of clipper ship.

Ocean shipping requires that we make a trade-off between speed and capacity. A longstanding compromise had been struck by 1845. Cargo moved over the seas in slow-moving, high-capacity merchantmen. That balance was briefly upset soon after San Francisco became the golden gate to Western America. It became a profitable port even before gold was found in 1848.

The same booming economy that drove shipping to California also drove the market for Chinese tea. As shipping rates rose from $10 to $60 per ton of cargo, it suddenly became profitable to build and operate ships that functioned more like racing vessels than traditional cargo carriers.

The term "clipper ship" was taken from the fast little Baltimore Clipper design, which dated back to the War of 1812. The biggest of these ships, the *Ann McKim*, weighed less than five hundred tons. But in 1843, it traveled from Canton, China, to New York with a lucrative cargo of tea in only 96 days. The implications were clear to merchants.

Masts rose into the sky. Hulls developed a knife-edged bow. The widest beam was moved over halfway back from the bow, giving the ship a

(From *The Story of the Ship*, 1919)[2]

fine, efficient, hydrodynamic shape, along with greatly reduced carrying capacity. Economy and long life were literally thrown to the winds. The focus on speed begot ships that seemed to have sailed out of a child's dream.

Clippers were tall and beautiful. A third of an acre of canvas would drive one at 14 knots. For a while those expensive carriers paid for themselves on a single voyage. When Nicholas Dean tells the story of the clipper ships, he calls them the SSTs (supersonic transports) of the nineteenth century—speed at any cost.[3] Their beauty moves us when we see them in pictures, for they touch our minds the same way they touched the nineteenth-century mind.

But we can see them today only in pictures. The financial boom ended in 1855. After that the fast ships vanished. The last one afloat was serving as a barge when it accidentally burned in 1923, and none were among the tall ships that sailed into New York Harbor for the United States bicentennial celebration in 1976.

The new steamships, by the way, were not what ended the day of the clipper ship. Steam packets had been in operation for about two decades before the tea boom. And the clipper ship had long since come and gone before steam finally drove most of the purely sail-driven merchantmen off the seas in the late nineteenth century. As a flow of goods was established, speed was no longer worth the load-carrying inefficiency of these glorious sailing vessels.

We see here that, just as the term *technology* might well be stretched to include physical means as well as machines, the term *infrastructure* must also be stretched to include economic circumstances. In this case, the practical constraints that bound shipbuilding were momentarily released.

Whenever that happens, the results may be artificial, but they can be stunning to watch until the natural constraints reassert themselves.

Our other example works in much the same way as the clipper ship. This was Alfred Ely Beach's invention of the pneumatic subway.[4] Beach was upper crust, educated at Yale, and given management of the New York *Sun* newspaper by his father in 1848. He was only 22 when he took over the *Sun*'s offices in the Wall Street area.

Manhattan was, by then, turning into a high-population-density nightmare, and Beach was a young man with ideas. Two years before, he had bought the one-year-old *Scientific American* magazine, and he was on the way to bringing it to greatness. He had also opened a major patent agency that would eventually deal with clients like Morse, Edison, Bell, and Ericcson.

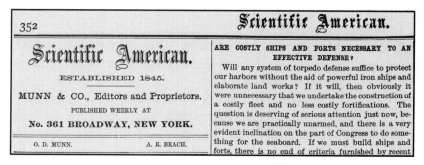

Scientific American's masthead for June 6, 1885, shows Beach serving as co-editor. Whether or not he wrote the editorial on the right, it is nevertheless highly relevant to the discussion of battleships in this chapter.

All the while, Beach watched Manhattan's streets growing hopelessly befouled and dangerous as four or five horse-drawn omnibuses might rattle by each minute. He thus turned his attention to the idea of urban rail service. The British had just built a small experimental subway line in London, but it would be another 30 years before electric-powered subway service was finally established in America.

In 1866, Beach petitioned the City for something called a "postal dispatch charter." This was actually a smokescreen, a way to get authorization to build a subway system without letting the City of New York know what he was doing. He meant to start with a small, three-hundred-foot demonstration line under Broadway. He had to keep it secret from the corrupt Boss Tweed (of Tammany Hall infamy), because he knew Tweed would extort extra money before letting him dig.

To cut the tunnel, Beach improved on the hydraulic shield designed by the French-born British engineer Marc Brunel. Brunel had begun the first tunnel under the Thames River in London, and his son Isambard

Kingdom Brunel finished it. Like Marc Brunel, Beach put his own son in charge of cutting their nine-foot-diameter hole. Father and son worked in secret at night, trucking away dirt in wagons with muffled wheels. Before America finally committed itself to serious subway development, the British had already turned about and used *Beach's* design in *their* tunneling.

The finished system had a single pneumatically driven car that shuttled people between two sumptuous stations, each decorated with paintings, frescoes, and fine furniture. It had cost Beach $350,000. When the subway opened, Boss Tweed was at first flabbergasted, then enraged. He managed to close Beach down within a year; but, not long after that, the great political cartoonist Thomas Nast exposed Tweed.[5] Tweed was indicted for graft, and Beach's charter was reinstated.

Soon after, a stock-market collapse put Beach out of business for good. His subway was sealed up and forgotten until 1912, when workers digging an extension of the BMT subway line suddenly broke through into Beach's old tunnel. It could have been King Tut's tomb with all that sealed-up elegance—the old rail car, the décor, the hydraulic shield. Today, Beach's tube is part of the BMT line.

Beach was a dreamer who promoted invention, fostered change, pioneered the art of tunneling, and helped establish the venerable *Scientific American* magazine. His subway was a clear case of too little too soon. Subways were a straightforward extrapolation of railroads, but they could come into play only after a much larger construction infrastructure had

This cartoon by Thomas Nast shows Boss Tweed as "Our modern falstaff reviewing his army" of thugs. (From *Th. Nast: His Period and His Pictures*, 1904)

been developed—after more had been learned about building with iron and steel on a large scale.

And here we reach another dimension of short-lived technologies. It is that a new technological means can drive people to invention—often without their seeing the whole picture. Steam was surely the motive *force* behind the industrial nineteenth century, and it drove minds to create all kinds of new uses for it. But iron was the underlying *substance* of this new world. Iron is the fourth most common element on Earth (even our blood is part iron). And an increased fascination with iron was destined to drive many miscalculations of the technological future.

In 1709, James Darby's use of coke to smelt iron had led to the widespread commercial use of cast iron. Darby created the Coalbrookdale Ironworks. Then, in 1777, the first delicate iron bridge reached out to span the Severn River at Coalbrookdale. England went on to build iron bridges and iron furniture—iron machines and iron ships! The wonderful plasticity of cast iron allows one to cast complex forms cheaply and easily. Then the Bessemer smelting process, only 11 years old when Beach began his subway, made iron even more plentiful in the form we call *steel*. Iron so fed and formed new dreams and new excesses that, in 1884, Swinburne wrote:[6]

> Not with dreams,
> but with blood and with iron,
> Shall a nation be moulded at last . . .

The Bessemer Process, from *Great Industries of Great Britain*.[7]

I suppose Swinburne was thinking of weapons when he used the word *iron;* but, by 1884, England had quite literally been molded in iron for a century.

America meanwhile relied upon on its huge reserves of virgin wood, a luxury that deforested Europe could not enjoy. We therefore moved less rapidly toward the use of iron. But we craved something else that Europe had and we did not. That was elegant masonry. We had been rendering the classical stone forms in wood from the beginning.

Then Philadelphia architect John Haviland took the idea of imitating stone with other materials a step further.[8] In 1828, he was given the task of designing the Pottsville, Pennsylvania, Miners' Bank. Since there was no local quarry for stone, he cast an iron façade in the shape of stone columns and cornices. In 1848, another builder, John Bogardus, began doing that on a much larger scale in downtown New York. Soon Bogardus was prefabricating and shipping iron facing all over America. A cast-iron wall, with wide window spaces to admit light, was still very strong.

And, if casting simple classical columns was possible, why stop there! For a season, iron facings became wild parodies of classical elegance. Makers next began producing iron walls that would bear the weight of a building.

But then the worm of doubt crept in. People began to see that painted iron is not really stone. John Ruskin raised his voice against iron: "You can't have art where you have smoke," he cried. "[D]arkness . . . broods over the blast of the furnace." Then, in 1871, a terrible fire swept over the iron façades of Chicago (more on that in chapter 6). When the wreckage cooled, developers rebuilt.

America had, by then, begun manufacturing high-quality steel on a large scale, and Otis had invented his new elevator. A new breed of engineers began using steel girders to build far taller buildings. And the new esthetics of modernism was about to doom the frilly ornamen-

Iron front on the Greenleve, Block & Co. Building in Galveston, Texas, 1882. (Photo by John Lienhard)

tation that festooned most cast ironwork. Even as Swinburne wrote, architects were moving away from the old cast-iron buildings.

This 40-year moment was one of the shorter major architectural movements we've seen. Steel-girder construction was the replacement technology that would blow it away (more on that, as well, in chapter 6). Yet the old iron-faced buildings were robust. Many may still be found scattered about downtown Galveston, Texas. In the SoHo neighborhood of Manhattan, one can walk for blocks seeing hardly anything *but* cast-iron facades—now housing art galleries and coffee shops.[9]

Of course, Swinburne didn't really get it right when he said that a nation was molded, not in dreams, but in blood and iron. For dreams certainly drove the use of iron, and those dreams may have been most excessive when it came to matters of blood and iron. Perhaps you know an old nursery rhyme by Charles E. Carryl (still much repeated when I was a child):

> A capital ship
> For an ocean trip
> Was the Walloping Window Blind.
> No wind that blew
> Dismayed her crew,
> Or troubled the captain's mind.[10]

That calls up another technology doomed by its replacements to live, if not a short life, a seriously limited one. Strictly speaking, the term *capital ship* referred to any major warship. It had first been applied to the huge eighteenth-century wooden ships of the line. But it was soon applied to the new steel battleship (which was the linear extrapolation of the ship of the line). When we sang this old song, we still regarded battleships as the grandest machines afloat, even though their weaknesses were finally coming to light.

All the major nations built battleships. Some weighed over 70,000 tons. In 1906, the British named one new battleship *Dreadnought*—a name that

HMS *Dreadnought*, as pictured in the 1911 *Encyclopaedia Britannica*.

was soon applied to a whole class of major battleships. Surely such an engine of war had naught to dread from anyone. As many as nine of the big guns on such a ship could be as large as 18 inches in diameter.

Our view of the big battleship changes when we read historian Garcia Rodriguez y Robertson's description of them.[11] They have joined only a few great battles. In 1898, the U.S. Navy scored a one-sided victory over the Spanish at Santiago, Cuba. It destroyed a Spanish squadron with no loss of American life, and the battleship was hero of the day.

But records of shots fired and hits scored tell a different story. The Americans fired 6,000 rounds. Only 130 of those hit *anything,* and only 2 hits out of the 6,000 rounds came from big battleship guns. Far from sinking anything, those 2 did only topside damage. Similar stories emerge from other naval artillery duels.

My 1970 *Encyclopaedia Britannica* claims that those big guns could hit a target with great precision at 25 miles. They could not. Consider the problem: To do damage, a shell must strike perpendicular to a ship's armor. A big shell coming in 20 degrees off perpendicular won't penetrate the nine-inch armor protecting most battleships at the water line. Long-distance shells come down in an arc. The chances of hitting anything become vanishingly small, and shells necessarily deal glancing blows when they do hit—unless a lucky hit should land on an enemy deck.

Battleships *are* vulnerable to *torpedoes*, however (note the almost prescient remarks in the right-hand side of the *Scientific American* masthead image shown earlier). When a torpedo explodes outside the hull, below the waterline, it does not diffuse a portion of the blast into surrounding air, for there is no air. The water instead focuses the explosion into the hull. A quarter-million tons of battleships were lost to torpedoes in World War I. And battleships did more damage blundering into friendly small craft than they did shooting at enemy torpedo boats.

In 1921, British and American representatives met in Washington, where they agreed to call a moratorium on capital-ship building. The traditional military cried out its objections. Later, that conference was blamed for leaving America unready in 1941. In fact, what it had done was to turn the Navy's attention to aircraft carriers and submarines. The conference may have been what saved America in the war.

So was the "capital ship on an ocean trip" really just a child's dream? On the one hand, we certainly do love to imagine sitting off at a safe distance and pounding an enemy into submission. In all its many forms, that dream has repeatedly turned out to be little more than the sound of the walloping (or beating) window blind. However, the battleship may well have served one real tactical function over the years, and that was the show of strength—a means for frightening a lesser enemy.

This 1919 image from *The Story of the Ship* bears the revealing caption, "The perfection of ship building, the USS *New York*." The most telling feature of the illustration is the dirigible lurking in the sky above—portending the future.

The clipper ship outran its economic "infrastructure" within ten years, but the battleship lingered much longer. It was buoyed, not only by eighteenth- and nineteenth-century expectations of naval warfare, but also by a few one-sided victories that contributed to its symbolic value. In any case, we did not give up our expectations of the battleship until submarines and airplanes had been in use for many decades.

Other heroic nineteenth-century steel structures almost reached completion but, without the kind of expectations that surrounded the battleship, were stopped before they did. A classic case was a bridge designed by the great bridge-builder John Roebling—a bridge so bizarre that not even he could bring it to reality.

This bridge was to have lain right next to the location of the late-twentieth-century Three Rivers Stadium—home to the Pittsburgh Pirates, and to the Steelers during their legendary years. The stadium was named for its location, just north of the point where the Allegheny and Monongahela Rivers meet to form the Ohio. The rivers form a letter **Y** centered on Pittsburgh. When Pittsburgh was mushrooming into one of America's major industrial cities, two centuries ago, bridges were a major concern. (Today, many bridges and tunnels necessarily crisscross the three rivers.)

Roebling was a young Prussian engineer when he came to America in 1831 to help form a utopian community north of Pittsburgh. Ten years later he gave up that agrarian life and went to work developing wound-steel cables for pulling barges. He then realized he could revolutionize bridges if he used his cables to suspend them.

Roebling began building bridges in Pittsburgh. He built an aqueduct suspended across the Allegheny and a 1,500-foot bridge over the Monongahela. He bridged Niagara Falls, and he eventually gave us his masterpiece, the Brooklyn Bridge.

But Roebling's Tripartite Bridge was never built.[12] The Tripartite Bridge was to have formed an inverted **Y** right where the three rivers meet. The

Roebling's cable work on Brooklyn Bridge tower. (Photo by John Lienhard)

three branches of the bridge, radiating from a central point in the water, were to link the three parts of the city. Each branch was a single span about a quarter-mile long. The Pennsylvania governor okayed the bridge in 1846, and stock went on sale.

Then competing forces rose up against the bridge. It would put ferry-boat operators out of business. Some businessmen saw it as a threat to real-estate values in the city center. People were told that it would block river traffic and take cash away from other important projects. In the end, no stock was sold, and the project was put on hold.

The bridge offered one really serious stress design problem. Suspension-bridge cables carry huge loads. They're usually anchored solidly into the earth. But in this bridge, one end of each leg had to anchor at a central pier out in the water. Cable forces from three directions would have to balance *one another* at that pier.

Could Roebling have done it? He might well have done so, but he died before the project was resurrected in 1871. This time, developers turned to Roebling's son Washington, who was then finishing the Brooklyn Bridge, and gave up on John Roebling's cable-balancing act. They opted for safety and proposed to make a solid truss bridge of one branch. That way, the other two suspension branches could anchor their cables on it. But people still wouldn't invest.

The project died a second time, and we are left with *what ifs*. The tip of Manhattan is much like Pittsburgh, with a river on either side. We can imagine traveling over a three-way bridge connecting Manhattan, Brooklyn, and New Jersey. Roebling had repeatedly run to the fringes of supporting infrastructure and gotten away with it. The Tripartite Bridge was one case in which he went just a hair too far and brought a political backlash down upon himself.

Steam and steel met the new force of electricity head on in another brief technology, one that symbolizes San Francisco as powerfully as the Eiffel Tower symbolizes Paris. The *cable car* was a clever answer to a nasty problem. But it was one more extrapolation that was almost immediately replaced.

In 1869, neither electricity nor internal combustion engines were in general use, and the only obvious way to power streetcars (or *omnibuses*, as they were more often called) was with horses or a steam engine. At the time, you could hardly put a fleet of mini-steam locomotives on city streets. Thirty years later, the steam car would emerge as a head-to-head competitor with the new internal combustion engines, but for the moment, horses were the only viable means for pulling buses.

Then, in 1869, an inventor and developer named Andrew Hallidie visited San Francisco. There he chanced to see an accident. He watched four horses pulling a streetcar up one of the city's murderous hills. One of the horses slipped, fell down, and started a chain reaction. The rest of the horses fell, and the car rolled down the hill dragging the panicked beasts behind it.[13] Horses might be well-suited to pulling buses in other cities, but hilly San Francisco obviously posed a special problem.

Since Hallidie was, at the time, a wire-rope salesman, he saw an alternative. One could cut a slot in the street and then run a moving steel cable through the slot. One could use a stationary steam engine to drive that long moving cable. The cars would be equipped with a grasping device that the operator could drop down into the slot. To move the car forward, the operator would grasp the cable; to stop, he would release it.[14]

Hallidie's idea was not entirely new. Something like it had been tried in London and again in New Orleans. But Hallidie organized a company, he solved the thorny problem of making a really good quick-release mechanism to grip the cable, and his new system made its debut on precipitous Clay Street in 1873.

Before long, six hundred streetcars were riding a network of inch-and-a-quarter cables. The length of some cables reached four miles, and they towed streetcars all over San Francisco at a steady nine miles per hour.

Of course, cable car systems worked equally well on flat streets. They quickly spread across America, but they did so just as Edison and Westinghouse were coming out with electric motors and public electric-supply systems. By 1890, cable cars had, in the blink of an eye, been made obsolete by the electric streetcar.

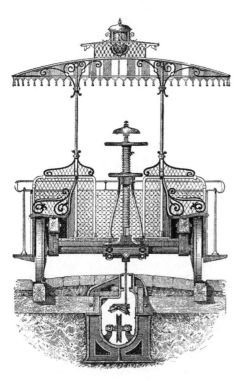

Cable railroad illustration, from the 1892 *Appleton's Encyclopaedia of Applied Mechanics.*

Most of the San Francisco cable lines were abandoned. Since Hallidie had conceived those proud little vehicles with a special capacity for scaling San Francisco's precipitous hills, a few of that city's cable lines survived. They survived the earthquake. They survived the fire. And they're still running. Of course, the system has shrunk. It had 110 miles of track in its heyday; now it has fewer than 11.

What remains of it is a piece of Americana, frozen in time. Sentiment has made the cable car into an icon, for it (like the Pony Express or iron building facings) really was the right technology in the right place. But it could only remain right until a radical new technology, already waiting in the wings, emerged to change everything.

The most elementary conditions for the survival of a new technology are that it must arrive at the right time and satisfy the true wants and needs of a people. It must also be sustainable, not just by infrastructure or the lack of any better replacement, but by expectations that either exist or can reasonably be created.

This last point is tricky and must be hedged, for in rare cases a technology will overcome a lack of infrastructure and expectation. In rare cases it will leap across the divide and find footing in some new future whose time has come. Let us look at one more example of a short-lived technology, which, I believe, illustrates this point by its profound inability to make such a leap.

The little-known West Baden Springs Hotel was extrapolated from an existing technology to create grandeur on whole new scale, only to be left utterly behind by a future that went off in a different direction.[15]

By way of comparison, consider the Houston Astrodome. When it opened in 1965, its 680-foot-diameter dome was the largest in the world. The Astrodome served into the twenty-first century, having been upgraded many times. It was the right technology in the right place at the right time, and vast domed spaces like the Astrodome became commonplace after 1965.

No dome had, by the late nineteenth century, exceeded the free span of the ancient Pantheon, St. Peter's Church in Rome, or Brunelleschi's Dome (each of which is only about 140 feet across). Then, in 1888, the railroad reached the remote corner of southern Indiana that is West Baden Springs. A businessman, Colonel Lee Sinclair, had made a great deal of money. Now, with access to rail service, he decided to create a spa in West Baden Springs, which had been named after the local sulfur hot springs.

Sinclair built a five-hundred-room hotel, a casino, a Catholic church, an opera house, a gymnasium, a swimming pool, and more. His brochure listed 50 ills that the waters would cure. They included alcoholism, pimples, gallstones, and other ailments. The railroad hauled in visitors from the East, and business boomed until 1901, when a terrible fire destroyed all those wooden structures in an hour's time.

The owner of Sinclair's nearby rival, French Lick Springs, seized the moment to announce a big expansion. The battle of the spas was on. Sinclair immediately commissioned a great fireproof pleasure palace with over seven hundred rooms, many of which would surround and face into a vast domed atrium. The contract stipulated that it would be finished within two hundred days. To accommodate thermal expansion, the inverted-bowl-shaped dome of structural steel was to ride on rollers. The rollers sat on the flat tops of six-story columns. This dome was two hundred feet in diameter. The first guests entered this monumental hotel one year to the day after the fire.

As modern medicine eroded the plausibility of the spa's miracle cures, the emphasis had to shift from the waters to the recreational facilities. By the time of the Depression, small handfuls of the surviving wealthy wandered its cavernous immensities feeling more like they were in a mausoleum than a resort. The hotel closed in 1932 and never reopened.

The Jesuits finally bought it for the sum of one dollar. They occupied it until 1964. Then it became too much for them as well. Various people owned it, and tried to occupy it, but finally it fell into ruin. One outer section collapsed. In 1996, the Historic Landmarks Foundation bought it for a quarter of a million dollars. At this writing, the foundation (with corporate help) is trying to restore it and to find a permanent owner and use for it. It meanwhile runs tours for ten dollars.

Yet that vast interior remained close to the largest in the world for two-thirds of the twentieth century. Sinclair's luxury hotel was so grand, so pretentious, and so wildly beyond any human purpose that it overstated what America was, even in a grand age. The West Baden Springs Hotel did mirror turn-of-the century American expectations, but it went too far beyond them. In the end, it outran what I would like to call the infrastructure of expectation.

And so we powered our way up to the new century, trying out ideas, but only faintly sensing the truly incomprehensible change that was moving in upon us. Of all the forces about to rise up in our face, only electricity was evident. We knew perfectly well that it was poised to disrupt everything, just as it had aborted the Pony Express and the new cable car systems. The telegraph was well established, the telephone was just getting the attention of consumers, and public electrical supply systems had come on the scene only 18 years before.

Those technologies were obviously just hints of the eventual impact of this mystical new force, and writers were struggling to see what results it might one day have. One of those was Lyman Frank Baum, who wrote *The Wizard of Oz* in 1900. The Oz series constantly reveals Baum's fascination with new and emerging forces of nature, and he helps us to see what perplexity we faced at that point.

Baum was born in upstate New York in 1856. His father, a barrel maker, went on to become oil rich. Baum was dogged by poor health and driven by a fertile imagination. At 17 he started a newspaper. At 24 he became manager of several theaters owned by his father. Baum wrote and mounted musical plays. He ran a newspaper in South Dakota. His first book was about raising chickens. Then, in 1900, he wrote the first of 14 or more Oz books. They are wonderful stories. (I know; I was raised on them.)

But the Oz books were only a fraction of Baum's voluminous writings for children. A year after *The Wizard of Oz* he wrote a little-known book, *The Master Key*.[16] It tells of Rob, a boy whose father allows him great freedom to do electrical experiments in his workshop. Rob plays with telephones and electric motors.

Rob's mother worries about his safety, but his father replies, "Electricity is destined to become the motive power of the world." So Rob strings wires and throws switches. One day he connects two wires, unsure of what he is doing, and a genie appears in a great flash of light. Rob has inadvertently summoned the Demon of Electricity by touching the *master key of electricity*.

The demon promises to give Rob three electric items from the future, each week, for three weeks. He tells Rob to use the devices with care and circumspection and to show them to the wisest people in the land. The first three are a tube that stuns an enemy for an hour without killing him, a wrist device that transports one through the air, and a box of pills that serve in place of food. Rob takes them, activates the wrist device, and flies off. His rashness leads him into terrible dangers; but he returns safely to receive the second three devices.

This time the demon gives him a pair of glasses that reveal a person's character, a sort of hand-held TV screen that shows what's taking place anywhere on earth, and a garment that wards off any assault. Rob barely makes it back from the adventures of the second trip. This time he returns chastened and much wiser.

When the demon offers the third set of devices, Rob says, "Keep it." The horrified demon says, "You might have hastened [a new day] if you'd been wise enough to use your powers properly." "But I'm NOT wise enough," Rob cries. It's up to humans to *become* wise enough first.

That sounds a little like Dorothy deciding that being back in Kansas is best after all. Baum reminds us that Rob, Dorothy, and all of us really did stand at the portals of Oz in 1900. As the Demon of Electricity went *poof*, an army of smart people stepped up to begin creating the same wonders that Rob refused to accept.

I am not inclined to credit Baum with great insight here. What he wrote is more important for the way in which it caught the temper of the times. Baum was imaginative, but he had trouble escaping much of the thinking that surrounded him. He was, for example, a great deal more racist than even the racist world he lived in.[17] What he did in this little book was to leave us, not with a vision of electricity-to-come, but rather with a story about the foreboding that touched people a century ago. He spoke to the confusion we faced in 1900.

A century later we've seen some of the demon's gifts made into near reality, and we've thought about ways to bring some of the others to fruition. But in 1900 people had caught just a glimpse of that demon, and they had no idea where he actually meant to take them. I said at the outset that the new sciences would be at the root of the new twentieth-century world.

Baum accurately singles out the emergent phenomenon of electricity as the great agent of change. It is surely high among the seemingly mysterious new forces—electricity, X-rays, radiation, and radium—that we must examine next as we search for the invention of *Modern*.

Still, *Modern* did not, and could not, rise from extensions of earlier technologies. It rose, instead, out of a complete intellectual disruption. Historian Henry Adams, great-grandson of President John Adams and grandson of President John Quincy Adams, visited the Paris Exhibition of 1900. When he wrote about the experience in 1905, he saw that disruption with disturbing clarity. He looked at human contrivances all around him and realized that they only hinted at the real nature of the coming twentieth century. What he actually wrote was this: "[I found myself] lying in the Gallery of Machines, my historical neck broken by the sudden irruption of forces totally new."[18] It is to the nature of that irruption that we need to turn our attention next.

3

"The Irruption of Forces Totally New"

S ometimes it helps us to understand the past if we identify an *annus mirabilis*, a year of wondrous movement and change. One such year is 1776, when Americans declared their independence from Great Britain. The year 1776 was also when Boulton and Watt went into business making steam engines, thus declaring a kind of economic independence for British commoners. And it was the year that Adam Smith published *The Wealth of Nations*, which repudiated the old mercantile economics of imperialism.

The world did not turn upside down precisely in 1776. Many of the important events that fix 1776 in our minds occurred before that year and after it. But 1776 was roughly when eighteenth-century revolution truly gelled and it gives us a good place to hang our hats when we think about that epoch.

If I had to identify the next comparable *annus mirabilis* after 1776, I might reach for 1901, the year just after Henry Adams faced the "sudden irruption of forces" that would shape the new century. The world was shifting once more, and in 1901, the irruption found a very specific focus in the publication of a paper by Max Planck.

Planck had finally solved a longstanding problem of physics when he correctly predicted how the energy emitted by a blackbody is distributed among the different wavelengths of light, heat, and other forms of radiation. (A blackbody is an ideal surface that absorbs all light and heat radiation reaching it, reflects nothing, and perfectly emits radiation.)

No one had managed to do that until Planck abandoned the commonsense idea that energy is a continuous and infinitely divisible quantity. He tried assuming that energy flows in discrete packets, which he called *quanta*. He had meant to predict the energy distribution by allowing the size of his quanta to approach zero. To his surprise, the physical constraints on the energy distribution required his quanta to retain a finite size, yet the result was a completely accurate prediction.[1] It would take another quarter century for the full meaning of that unexpected result to become clear.

One might well be tempted to look upon Planck as a lucky lottery winner, but what he did was no more blind luck than James Watt's invention of the external steam engine condenser. Both were extremely bright and insightful people. Both were wonderfully attuned to the zeitgeist that was forming around them. Neither knew the full extent of change that would follow, yet each had managed to find the precise crest of the wave rising beneath him.

After Planck's paper was published, the very foundations of a reality that had seemed as plain as the nose on our face began slipping away. Planck had begun the process that would turn photons from material elements into what more closely resembled a new essence of the old alchemy. The hard-edged reality of the late nineteenth century is one that we still try to claim, but shall never own again. Definiteness, predictability, and certainty began crumbling under our fingers.

By 1901, electricity had been the focus of great attention for a century and a half. This new agent of change had already assumed a very large place in everyday life. It was quite clear that electricity would be the formative agent of the new century; telegraphs, electric motors, light bulbs, and telephones were already in use. The real fun began as we intensified our attempts to know just what electricity *was*. Now we began posing that question in ways that would have to end with the deconstruction of matter, energy, and all that we had ever tried to call *real*.

Another idea that had been in place for just 99 years in 1901 was John Dalton's fresh articulation of the old atomic theory of matter. Dalton's favorite recreation had been lawn bowling, and he formulated a theory of chemical reaction based on the notion that matter was made up of small spherical atoms. He believed those atoms to be miniature versions of the bowling balls he used every evening on the lawn. Dalton's atoms were very real.[2] He vigorously objected to the abstract chemical notation that arose in the wake of his theory—the notation we use today.

We typically write chemical equations that look like this:

$$2H_2 + O_2 \rightleftharpoons 2H_2O$$

In place of letters, Dalton demanded that we represent the different atoms with circles, each containing its own symbol. Atoms were no mere conceptual prop for Dalton, and he meant to give people no opportunity to take his rules of chemical combination as figurative rather than literal.

Some atoms in Dalton's notation.

 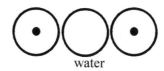

Two molecules in Dalton's notation. (Actually, Dalton did not yet have details of atomic weights fully worked out, and he pictured the water molecule with only one hydrogen atom.)

In that demand Dalton was very different from Planck, for Planck had avoided assigning explicit meanings to his quanta of energy. He had made energy into quanta only as a mathematical trick, without claiming to have made a conceptual shift. Dalton knew that he had done more than just *describe* reality. He understood that he'd *altered* reality as well.

Dalton's atomic theory of chemical combination became, at first, the basis for the new field of process chemistry. All the while, scientists allowed his atoms to hover between literal and figurative truth. In 1860, however, the brilliant young Scottish scientist James Clerk Maxwell and German scientist Ludwig Clausius (who had been a founder of classical thermodynamics) began to move far beyond mere questions of chemical composition.

They began showing exactly how moving atoms and molecules gave rise to the physical character of gases. They showed how to relate the pressure, volume, and temperature of gases to the movement of their atoms and molecules. They showed how molecular motions determined such specific features as viscosity and thermal conductivity. As they did so, atoms mutated from conceptual tools into the almost tactile beams and girders that give matter its physical attributes. While the logic behind the new kinetic theory of gases was hard to refute, some logical holes did open up on closer analysis.

The idea that matter is made of real particles, too small to see, continued to bother people. One problem, in particular, seemed to cast lingering doubts over Maxwell's generally scrupulous mathematics. It was the paradox of molecular reversibility. When molecules collide, they bounce off one another's force fields with no friction and no energy loss. If time ran backward, each collision would reverse itself perfectly.

That, however, sets up the following absurdity: First, suppose you open an air tank, and air molecules begin rushing out. Now suppose the motion of each molecule could somehow be reversed. Wouldn't history itself run in reverse? Wouldn't the molecules rush back into the tank? We've created a situation that sounds as silly as it sounds logical. The reason it sounds silly is that it would violate the second law of thermodynamics; and that is something that we all understand in the pit of our stomachs even if we have never studied it formally.

Time is directionless on the molecular level, where motion is perfectly reversible. Here in the visible world, on the level of human perception, the past is irreversible and can never be undone. Time's arrow flies from past to future, and air will never flow back into the tank. The idea that processes had to be reversible on the atomic level, and irreversible on the level of human experience, seemingly offered good cause for not taking atoms seriously.

The next great molecular theorist, Ludwig Boltzmann, went further than anyone before him, and the resistance of doubters rose proportionally. Boltzmann turned superb mathematics on the question of irreversibility. He showed how rules of averaging would never let such a reversal occur. In any large collection of molecules, disorder would continue to increase even if one *could* find a way to reverse the motions. The gas would continue to flow out.[3]

Yet something was missing. While Boltzmann's calculations rested upon probability, our minds keep going back to that single, perfectly reversible collision of two particles. Classical physicists who had resisted Boltzmann's molecular mechanisms could now go on the attack.

Boltzmann was brilliant, but he had a history of depression and mental illness. It caught up with him. He couldn't answer his critics, yet he knew he was right. He wrote, "[theory] fills my thought and action . . . no sacrifice for it is too much for me . . . [it is] the content of my whole life." On September 5, 1906, the 62-year-old Boltzmann slipped a noose around his neck and hanged himself. He almost certainly committed that irreversible act because he despaired of being understood on the matter of irreversibility.

Einstein and a new breed of physicists were starting to take Boltzmann very seriously by 1906. Had he waited just a little longer, he would have seen his genius triumph. The new science of quantum mechanics was already taking shape, and Heisenberg's Uncertainty Principle would eventually say that the real absurdity lay in supposing that motions could ever be perfectly reversed in the first place (more on the meaning of these ideas in a moment). In a quantum universe, Boltzmann's mathematics still makes sense. Only the idea of perfectly reversing time becomes unthinkable.

Boltzmann's proof that irreversibility flows from reversible molecular action included an explanation of the second law of thermodynamics (which says that spontaneous processes are always irreversible on our scale of perception). He showed how that essential fact of nature arises out of molecular considerations.

The new intellectual apparatus that Boltzmann had created was vast. He had finally built the bridges that would connect molecular motion with the world that you and I experience. Further, he knew what he had done; he directed that his tombstone have carved upon it his equation relating entropy (S) to molecular probability (Ω):

$$S = k \ln \Omega$$

where k is a number called Boltzmann's constant.

Just as it took quantum mechanics to resolve the paradox of irreversibility, it also took quantum mechanics to resolve another failure of Boltzmann's kinetic theory. The theory could not correctly predict the distribution of blackbody radiation. It remained for Planck to reveal that the chaos actually ran much deeper than our inability to know all the movements of atoms and molecules. No more tiny bowling balls! Atoms were indeed more abstract than that, but in ways that no nineteenth-century opponent of Dalton could have imagined.

The twentieth century thus began with revelations whose ramifications are still spinning out today. The people in the street knew nothing of Boltzmann or Planck in 1901, but they could see other straws in the wind—other new forces that were far more visible to the general public than the austerity of molecular/thermodynamic theory. Perhaps the most dramatic, even theatrical, of those was the discovery of X-rays.

Wilhelm Conrad Roentgen had been working with cathode ray tubes on November 8, 1895, when he suddenly discovered the presence of another kind of ray, which (unlike cathode rays) could "see" through certain materials. Roentgen called them X-rays because they were hitherto unknown and certainly not understood. The dramatic feature of the new X-rays was that they could pass through flesh, but not through bone.[4]

People suddenly had means for gazing right into the human body, and the story of

Wilhelm Conrad Roentgen (1845–1923), from *Light Visible and Invisible*, 1928.

Elizabeth Fleischmann (chapter 1) gives us an idea of how rapidly that fact reached the streets. The media tried to make a celebrity of the modest Roentgen, and he found himself trying to dissuade colleagues from calling his discovery "Roentgen Rays."

While X-rays rapidly did find many uses, they also demonstrate just how hungry and ready the public was for scientific revelations. Seldom has anything taken hold of the public imagination so powerfully and completely.

Nancy Knight tells how X-rays set the inventive muse in motion.[5] Right away, magazine cartoons celebrated the idea. A typical one showed a man using an X-ray viewer to see through a lady's hat at the theater. Some people were certain that everyone would soon own X-ray viewers, and they began to worry about privacy. Within a year of Roentgen's discovery, one company was actually manufacturing lead-lined women's underwear to protect them from such viewers.

X-raying became a new form of entertainment. X-rays were painless, so they had to be safe! We would X-ray ourselves for the sheer fun of seeing our bones. One might make an X-ray photo of one's hand and then frame it. X-ray mania swept the land, and we are left to wonder whether the following bit of doggerel, by a would-be poet, was entirely in jest:

Not worth your while
That false, sweet smile,
Which o'er your features plays:

Thy heart of steel
I can reveal
By my cathodic rays![5]

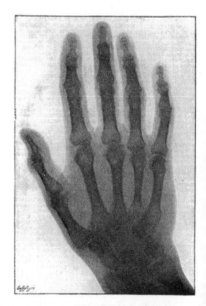

By the time I was a little boy, that idea had turned into stories about Superman's X-ray eyes. Meanwhile, real X-rays were cheap and accessible. I didn't have X-ray eyes, but I could X-ray my feet in new shoes at the department store. That particular technology didn't last long. By the 1930s, the terrible hazards presented by X-rays were becoming apparent.

After 1896, futurists seized upon the possibilities of seeing through opacity. That year, the president of Stanford University wrote about X-rays. He suggested that we might soon use them to read

Hand of the famous scientist Lord Kelvin, from *Light Visible and Invisible*, 1928.

thoughts. Whether that was satirical or serious, I don't know. But Edison was deadly serious when he set out to focus enough X-ray radiation on the human skull to watch the brain at work. He also proposed to heal tuberculosis with X-rays.

Today, every kind of radiation therapy and scanning device has found a medical use, and it's clear that some seemingly wacky ideas were actually headed in the right direction. Since X-rays attack human tissue, it was not at all far-fetched to expect that they might be made to attack germs or cancer. But, while people sought out potential benefits, they also minimized the increasing evidence of destructive side effects.

As a result, some early uses of X-rays were downright horrific. Historian Rebecca Herzig tells about their use in cosmetic hair removal.[6] Few of us today realize what serious business that became or that it lasted about 50 years. X-ray hair removal began right after Roentgen's discovery, and it continued through World War II.

X-rays simply made hair removal too easy. One early practitioner, Albert Geyser, created the Tricho machine. A woman would sit for four minutes with her chin in a holder while X-rays created a delicate ozone smell around her. The hair on her chin would be gone, and only later did she have to pay the terrible price. In 1954, long after Tricho machines were banned from use, one former patient wrote the following letter to a doctor (the spelling and grammar is hers): "[w]ithin the last few year's 'white spots' have appeared on my chin. This has been . . . heartbreaking to me . . . I have been wondering if ther might possibly be some new medical discovery which might help me."

By 1970, one-third of all radiation-induced cancer in women could be traced back to X-ray hair removal. The practice had been made illegal even before World War II, but it continued underground. Herzig begins her story by telling how San Francisco police staked out a house in 1940. Women entered the front door and left by the back. A doctor with a black bag came and went. What they thought was an illegal abortion mill turned out to be a secret X-ray hair-removal clinic.

Herzig suggests that X-rays came along when the time was ripe. Darwin's ideas were on everyone's mind, and that fed a whole new hair consciousness. Hair was what apes had, not humans. Suddenly, hair anywhere but on a woman's head was viewed as a medical disability. Yet the acceptable means of removal were all unpleasant. Hot wax and electrolysis were expensive and painful, and using a razor blade meant directly acknowledging the problem. X-rays, on the other hand, were painless and permanent. They were, in the modern vocabulary, *scientific*. Since X-rays, like tobacco, harbored a danger that lay far over the horizon, countless women continued using them in secret.

Two Canadian doctors finally gave a name to all the scarring, ulceration, cancer, and death that X-rays had caused. They called it the *North American Hiroshima maiden syndrome*. X-ray hair removal finally ended after 1946—after the public had seen all those terribly wounded Japanese women whose radiation hair removal had been no matter of choice.

To understand the public reaction to X-rays, it is important that we see beyond utility. Roentgen had provided the potential for medical diagnosis and radiation therapy. But X-rays had also initially promised means for doing what we have always craved to do—to see through the dark, to pierce the veil, to know the unknowable. In that, X-rays promised to endow us with magical powers. Small wonder they led us into excess.[7]

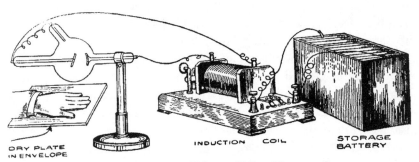

FIG. 281.—How to Take an X-Ray Photograph.

A 1925 book titled *The Boy Scientist* encourages boys to build and use their own X-ray apparatuses.

It is in the attempt to look inside the human body that we see how our vision often fragments just when it seems to become most penetrating. X-rays finally fulfilled this longstanding medical need. Medical historian Stanley Reiser tells how, a century before X-rays, a French doctor named René Laënnec tried to diagnose an obese young woman's heart disorder in 1816.[8]

Laënnec had tried thumping her chest, but she was too heavy; the sound told him nothing. The next obvious step would be to put an ear to her chest, but modesty forbade such intimacy. Suddenly, in a burst of either impulse or insight, Laënnec reached for a sheaf of papers. He rolled them into a tube, placed one end on her chest, put his ear to the other end, and was able to make out what was going on in her heart. On the spot, he had created the first stethoscope.

Three years later Laënnec published a book describing his design of a wooden stethoscope and its use. The idea caught on quickly. By the 1830s stethoscopes appeared with pliable rubber tubes, and binaural ones with earplugs followed. All the while debate raged—less over stethoscopes than over the more general question of what tactics were appropriate for diagnosis.

Laënnec's dilemma with that young patient would not have *been* a dilemma for most doctors around him. Thumping a thorax, or putting an ear to a heart, were radical forms of medical diagnosis in 1816. They would not have occurred to most of his colleagues. Diagnoses were usually made by looking at patients and listening as they reported symptoms. Doctors seldom questioned what patients said about themselves, and they inferred a great deal from outward appearances. Physical contact usually stopped at counting a pulse or touching a forehead.

Laënnec's ideas about thumping, feeling, and placing an ear to a patient went back to Hippocrates, who believed that all five senses should be used in diagnosis. An ancient Greek doctor might have diagnosed diabetes by tasting a patient's urine, but that kind of intimacy definitely did not appeal to eighteenth-century sensibilities.

Today stethoscopes let doctors keep their distance and still engage actual symptoms. This simple new instrument became the stalking-horse for a whole new kind of medicine—one in which doctors could bypass the patient's story and look inside the patient's body for direct evidence of disease. Stethoscopes were followed by ophthalmoscopes and laryngoscopes. For a time after X-rays were discovered, it seemed doctors had the noninvasive means they needed to skirt the patient completely.

This issue is now resurfacing as technology for reading the inside of the human body takes new leaps forward. Medical schools are finally returning to the idea that greater doctor/patient intimacy might be appropriate. The stethoscope once promised to bridge the gap and recreate doctor/patient contact, but, in the end, it gave means for standing even further away from patients. Any of us who have ever watched the movement of our own internal organs on a cool green computer screen feels the contradiction: that doctors may stand that close to our illness with eyes looking off at an image that feels no pain and suffers no anxiety.

Throughout the nineteenth century all kinds of reality came under increasingly close scrutiny. Our instruments increasingly led us, in a manner of speaking, toward the very atoms that made us up. Reiser uses the phrase "the lesion within" to describe the increasing focus of medicine upon specific seats of disease within the human body.

That focus had been developing since the late eighteenth century, and, when X-rays finally provided seemingly noninvasive means for looking straight into human tissue, doctors were empowered as they had never been. Medicine leapt ahead, but attention shifted away from the whole body. As X-rays illuminated the body's component parts, and our own account of our illness seemingly became irrelevant, medicine became a business between the doctor and our body. Our role as a patient was, for a season, reduced to that of a spectator.

A great fragmentation thus occurred after 1900. I have been asking what those "forces totally new" that irrupted around Henry Adams really were. You will notice that the implications of the question itself shift as we search out answers. No one makes that clearer than Adams who, in *The Education of Henry Adams*, portrays himself as a sort of everyman facing the juggernaut of twentieth-century technology.[9] In his chapter "The Dynamo and the Virgin," he tells how he was drawn back, day after day, to the 80,000 exhibits of the 1900 Paris Exhibition, trying to understand it all.

Adams's most important works of history were studies of Chartres Cathedral and the Abbey at Mont-St-Michel. He had come to see very clearly what a powerful social and technological force medieval Christianity had been. That force had found its focus in the person of the Virgin Mary. Now he gazed at wholly new technologies that had sprung into being in just a few years: dynamos, telephones, and automobiles, as well as the lurking inchoate forces of radiation and electric fields.

Adams realized that the dynamo would transform Western civilization just as powerfully as the Virgin had changed it in medieval Europe. By the time he wrote his book, the ideas that would create intellectual devastation—quantum mechanics and relativity theory—had already surfaced. While he doesn't mention them, he does, on a visceral level, see them thundering down the road. It was clear enough to Adams that Victorian confidence in science had outrun itself.

In the end, Adams lamented the blind spot that shrouded our vision of the change that was gathering momentum. The Virgin, he pointed out, had been the mystery that drove the medieval technological revolution. He says, "Symbol or energy, the Virgin had acted as the greatest force the Western world ever felt, and had drawn [our] activities to herself more strongly than any other power, natural or supernatural, had ever done." Adams understood that the dynamo and modern science were ultimately being shaped by forces no less mysterious. The great blind spot at the end of the nineteenth century had been a lingering and general *denial of mystery*.

As we began seeing the world anew in the atoms that made it up, mystery wove itself around those atoms. The new breed of twentieth-century geniuses was indeed *find-*

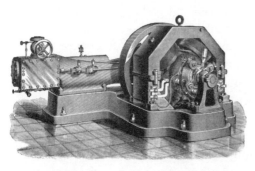

Multipolar dynamo from Adams's time, from *The Fireside University of Modern Invention*, 1902.[10]

ing mystery once again. We had tried to leave mystery behind in a rational, deterministic world. Now a great yawning pit of mystery opened up once more.

Four years after Planck's paper, in 1905, Einstein gave us the mysterious neverland of relativity. He led us into a thicket of counterintuition. Space now had to be curved, and the mass of an object had to increase with its speed. Mass and energy became two sides of the same coin. We seemed to be walking through a hall of mirrors.

In 1926, Schrödinger reduced the counterintuitive claims of quantum physics to a single mysterious hypothesis that we call the *Schrödinger equation.*[11] Physics books will often create the impression that we can derive Schrödinger's equation from other facts. We cannot. The equation is *itself* a new physical principle, constructed in such a way as to fit the seemingly bizarre set of facts that had emerged by the late 1920s.

When one solves Schrödinger's equation, one doesn't get the usual kind of information—velocities or positions. Since matter is no longer solid or definite, the solution is something called a *wave function.* Dalton's hard little bowling balls have turned into wavelike smears of *likelihood of existence*—what many people call "waves of probability." The words themselves make no sense until we deny all the perceptions of reality that have guided us all our lives.

One of the more disquieting ideas that emerged from Schrödinger's work was the way reality came to depend upon our observation of it. To explain that, Schrödinger offered the oddest thought model: he gave us a cat in a box.[12]

The riddle of the cat begins with Heisenberg's uncertainty idea. Heisenberg pointed out that the most precise measurement we could ever make would be to shoot one photon of light at a moving object, say an electron. But even so delicate a peek would change the position and motion that were being measured. At best, one must always measure with some uncertainty.

That's easy enough to understand. But an awesome subtlety turns it into a new tenet of scientific faith. It makes precise measurement unthinkable. That means we no longer have reason for thinking that the world *has* any ultimate precision to measure.

So we take the last terrible step. We admit that objects are all indeterminate and that electrons have fuzzy edges. When an electron collides with an object, it may bounce one way; it may bounce another. And, in that moment, Heisenberg made of uncertainty not a simple calculation, but another new physical principle—and a most disturbing one at that.

Schrödinger said if that is true, let us seal a cat, a Geiger counter, a fragment of radioactive material, and a bottle of poison gas into a box for one hour. We arrange things so the chances are 50-50 that radioactive decay will trigger the Geiger counter, activate a mechanism that breaks the bottle, and poison the cat. He then asks, "Will we find a live cat or a dead one when we open the box?"

That sounds like the "Lady or the Tiger," but it is much worse. The man who has to open one of two doors knows that a lady is behind one and a killer tiger behind the other. He does not know which door leads to the tiger, but the people who put it there do.

In this case we know exactly what is initially in the box. But radioactive decay occurs on the level of indeterminacy. No knowledge of the system in the box, however detailed, would ever allow anyone to predict the fate of Schrödinger's cat. One person will find a live cat; another might have found a dead one. Whether the cat lives or dies is absolutely unknowable until we open the box. *Two people in alternate universes, which were identical at the moment the box was sealed, might get different results.* The situation is worse than unknowable, for the answer has not been determined before the box is opened—before it is provided with an observer. To say otherwise would be to violate the Uncertainty Principle.

Physicists agonize while that Cheshire cat flickers on and off. They try to write wave functions for cats and gamma radiation. They conclude goofy things: maybe the cat in the unopened box is both alive and dead at the same time. In the end it is we who *make* the cat either alive or dead when we open the box. We determine the truth by making the observation. That makes an odd commentary on objective science, for now our subjective selves appear to be woven back into our science in ways we had never imagined they could be.

Strange realities thus began opening up as we gazed down into the granular structure of things. Perhaps William Blake's famous lines from "Auguries of Innocence" best describe what happened. A century before, he had written:

> To see a World in a Grain of Sand,
> And a Heaven in a Wild Flower,
> Hold Infinity in the palm of your hand,
> And eternity in an hour.

Actually, that understanding is much older than Blake. Medieval writers, contemplating the Incarnation, had much the same thing to say. Anyone familiar with Britten's *Ceremony of Carols* will recognize the idea surfacing within those ancient carols:

> Wolcum Yole . . .
> Wolcum . . . to more and lesse.

And again we hear,

> There is no rose of such vertu
> as is the rose that bare Jesu. . . .
> For in this rose conteinèd was
> heaven and earth in litel space, . . .

Blake reclaims the medieval notion that we find the infinities of our being by looking inward at small things. Henry Adams, historian of medieval Europe, looked at the coming of the twentieth century and saw that it would be fueled by the same vast mystery that had raised Chartres Cathedral almost eight hundred years earlier. An understanding, *totally new*, would well out of the grains of sand, out of the *litel space*. To get to *more*, we first needed to understand *lesse*.

And so we pursued the grains of sand. Artists may have led the movement. As early as 1884, Georges Seurat gave us his pointillist painting of bathers along the Seine River. In it, the detailed images emerge from individual points of color. Look at it closely, and all you will see are dots—atoms, grains of sand.

By 1912, Marcel Duchamp showed us how to fragment perception as well as representation when he painted his *Nude Descending a Staircase*. Suddenly we saw a cinematic series of pictures, not as a camera would record them, but broken into cubist fragments. It is as though we registered a nude suddenly coming down the stairs to join our gathering. Out of our embarrassment, we gaze at her only in a series of blinks, from the corner of one eye.

By the turn of the twentieth century, first the aging Liszt, then Skriabin and Debussy, had been trying to declare freedom from the tonal hierarchies that had bound music for five hundred years.[13] Music had, heretofore, been constructed upon scales that were hierarchical. The note *do* was called the *dominant* tone. *Re, mi, fa, sol,* and *la* each had its own hierarchical place. Finally, *ti* was called the *leading* tone because it led us back to the point of dominance—to the next *do*, an octave higher.

In 1908, Arnold Schönberg shed that hierarchy entirely in his second string quartet. There he went so far as to abandon key signatures, those signposts to the hierarchy of tones. Soon he and his students were writing new rules for giving each note equal importance. They shattered tonal hierarchies just as Schrödinger had destroyed the hierarchies of hard edges. And yet they showed that music could still be compelling, even after they had erased the class system that had made one note dominant over another.

James Joyce scattered the form of prose as readers had expected it to sound. It was not until 1922 that he wrote *Ulysses*, but Joyce placed his protagonist Leopold Bloom on the streets of Dublin during one day and night in 1904. Lost Odysseus—Ulysses—was plain to see in that sometimes bewildering sea of words, wandering the Aegean of his own Dublin.

I read *Ulysses* during the long hours of one night in 1955. I was the enlisted man in charge of the barracks while the rest of the company slept. Around sunup, I reached the point where Leopold Bloom returned at dawn to his wife Molly—his Penelope. As Bloom's long, muddled night ended, I finally realized that Joyce had reduced narrative to atoms of experience. In so doing, he had reclaimed the full intensity of storytelling. Joyce did exactly what Seurat had done when he reduced images to atoms of light while he maintained full control of the narrative power of a picture.

Thus the forces that irrupted were not electricity or X-rays. They were not pointillism, atonality, uncertainty, relativity, or any other one driving force. What irrupted was a totally new *whole* that welled out of a world now seen close up and fragmented. And the remarkable person who saw it coming was historian Henry Adams. I know of no one else who pinpointed anything so subtle within the materialistic ferment of the 1890s. I know of no one else who was keyed to see the mysterious reach of human need—who understood how ready we finally were to look, once more, for "heaven and earth in litel space."

4

A New Genus of Genius

s our view of reality fragmented, we began seeing mental brilliance as a new kind of commodity. Genius of a most unorthodox kind had clearly uncovered the "forces totally new" that were reshaping the world. Since genius had yielded such bizarre shifts in human perception, we began to see it as a developable asset. I recall people talking about genius when I was a child. What I.Q. rating defined it? Who qualified? One popular criterion was the ability to read and understand Einstein's Theory of Relativity. By the 1930s we had developed a kind of hyperawareness of genius.

Indeed, the word *genius* itself underwent a linguistic shift at the beginning of the twentieth century. For millennia, our genius had been our animating spirit—a kind of defining presence that makes us who we are. From that, it had mutated to mean a natural ability or disposition. Only in the new world of *Modern* did we begin using genius to identify that rare person who had remarkable mental capacity.

The formation of the Mensa organization in 1946 was an outcome of all that talk about genius. Mensa seeks members from the top 2 percent of the population, as identified by an appropriate intelligence test. (Interestingly enough, Mensans are smart enough to avoid using the divisive *term* "genius" anywhere on their web page.) But make no mistake: genius was on people's minds in the early twentieth century.

One genius in particular dramatically shows how we were beginning to think about the mind. William James Sidis was born in 1898 to Russian immigrants, intellectual refugees from the pogroms. Sidis's father Boris trained in psychology at Harvard. He worked with his son's namesake, the philosopher, psychologist, and writer William James, who had in turn studied with Louis Agassiz. Writer Amy Wallace tells how Sidis's mother Sarah gave up her own medical ambitions to join with Boris in forging intellectual greatness in their child.[1]

At first they appeared to be remarkably successful. Young Sidis could read at 18 months. He reportedly wrote four books and was fluent in eight languages before his eighth birthday. He spoke at a Harvard seminar when he was nine. Reports have it that he lectured on the fourth dimension, although that part may be apocryphal. He entered Harvard at eleven. William James Sidis certainly owned one of the most prodigious minds that ever was.

Still, he was not the only such person. At Harvard alone, he was merely the brightest of an amazing experimental group of prodigies who were to be found there in 1909. That group included the composer Roger Sessions and Norbert Wiener, father of cybernetics (we return to Wiener later).

Those awkward children suffered their isolated lives at a university that expected Eastern-finishing-school grace of its students. Sidis graduated at 16 and went to Rice University, where he served as a mathematics professor. Rice students ridiculed the childish Sidis for eight months. He finally gave up and went back to Harvard to study law.

Sidis went on to take up the socialist cause and was jailed in 1918 during a communist antiwar rally. It was in jail that he met the only woman he ever loved, an Irish socialist named Martha Foley. Meanwhile, the media hounded him. Sidis was determined to find privacy. He disavowed his knowledge of mathematics. The only work he would take was running calculating machines. He poured his energies into his hobby—collecting and organizing streetcar transfers. Sidis also wrote books, some under his own name, others under pseudonyms.

Most remarkable of these was *The Animate and the Inanimate*, published in 1925.[2] The book is heroic in its sweep. Among other things, Sidis predicts the existence of black holes 14 years before Chandrasekhar did. But primarily, he describes a cosmology that is explained by his recognition of the subjective character of the second law of thermodynamics (recall the discussion of Boltzmann in chapter 3).

Yet, despite its complex subject, not one equation—not one hint of Sidis's mathematical prowess—is to be found in the book. Sidis directly addresses Boltzmann's problem of molecular reversibility when he writes, "The second law of thermodynamics is really a mental law indicating the direction of the illusory flow of time. Time itself really exists as a two-direction affair, and really has no more flow than space."

Reading that, I feel the hovering presence of Sidis's deep-seated problems with human subjectivity. By the time he wrote *The Animate and the Inanimate*, he had fled his childhood, and he had fled his parents. When he died of a cerebral hemorrhage in 1944, he was still carrying Martha

Foley's picture. She had long since married someone else, but that didn't matter. Sidis could love only with his head. All his life he had vigorously rejected sex, art, music, and anything else that meant contact with the unwelcoming world outside his mind.

Wallace expresses her own anguish over that. She concludes her poignant biography of Sidis with a kind of cri de coeur, "Let us hope that [future gifted] children will grow up in a world that, instead of shunning them as oddities, will welcome and nurture their talents, . . . and their vision." While William James Sidis was neither the first nor the last child wounded by parents who thought they could open doors that no one else could, he powerfully embodies the remarkable way in which we went after genius, once we had seen it in the light of the new century.

Sidis's classmate Norbert Wiener was also the product of his parents' aim to create a mental giant. He entered Tufts University when he was ten, and he received his Ph.D. from Harvard at 19. For 41 years he taught at MIT. There he studied the analogy between human brains and machines. Having been driven throughout his childhood by ambitious parents, Wiener turned his attention to robotics and the machine-like aspects of the mind.[3]

He first created the word "cybernetics" from the Greek word for the steersman of a boat κυβερνητησ (kybernetes). A steersman sits with a hand on the tiller and sights over the prow to see where the boat is headed—toward a pier, for example. If the boat veers a little to the left, he turns it slightly to the right. When he overshoots to the right, he corrects it to the left. And so on.

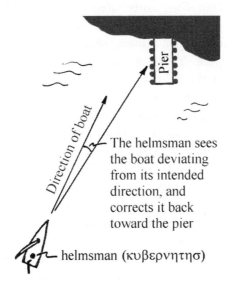

The helmsman sees the boat deviating from its intended direction, and corrects it back toward the pier

helmsman (κυβερνητησ)

The steersman constantly compares the actual direction with the intended direction and applies a negative correction, one that opposes the error. That's called *negative feedback*, and it is what all automatic control devices do. That's the way the level control in your toilet tank, your thermostat, and the ignition system in your car all work.[4] The Romans picked up the Greek word κυβερνητησ and transmuted it into *gubernare*, from which we take our word "governor." Governor has become another term that we have appropriated for a feedback control device.

Wiener studied the mathematics of the feedback process and went on to deal with what we might call the psychology of robots. He found, for example, that he could create behavior in robots that mimicked insanity in humans when he controlled them badly.

Psychologists have taken the term "positive feedback" from Weiner's work, but they misuse it. True positive feedback would be unstable and destructive—like a parent saying to a child, "Oh, Billy, you ran into the street again! Have an M&M." Technically, positive feedback is action that reinforces what it seeks to correct. What psychologists really mean is either positive *reinforcement* or negative feedback applied with good sense and courtesy. Little Norbert Wiener received plenty of feedback as a child. However, it appears to have been feedback lacking in the courtesy that parents owe their children.

Later in life, Wiener wrote books on the implications of the mind/machine analogy. He devoted his life to getting beyond the machine to find the human—to transcending his own overly controlled childhood. The title of one of those books is powerfully expressive of that impulse: *The Human Use of Human Beings*.[5]

This new obsession with genius traced, in one way or another, to three quintessential geniuses who were on everyone's mind in the early twentieth century. Thomas Alva Edison, Nikola Tesla, and Albert Einstein had captured the public's imagination with their accomplishments, and it helps to see them in relation to one another. Together, they show how the expectation for genius was mutating. Edison was entirely a creature of the nineteenth century. Einstein, on the other hand, embodied the twentieth century and everything modern. Tesla's genius lay in the way he rode the cusp between the two epochs.

The difference between Edison and Einstein was nothing so simple as invention versus science. Indeed, both contributed to science; both patented inventions. The difference revolved instead around a characteristic that the Romantic poets had articulated a century earlier. They told us, over and over, in many ways, that we create nature by dreaming nature—that we must look inside the human mind to discover reality. That, of course, is what we began doing on a grand scale in the twentieth century. The means that each one used clearly display the dividing line between the nineteenth and twentieth century.

I can best explain my meaning here by describing the odd way two threads converged rather late in my own life. The first thread was a small thing, a passing impression. During World War II, I watched one of the many movies about the evil empire of the Third Reich. In this one, the beleaguered hero, an anti-Nazi German, quoted Goethe to the heroine. The line, "Linger a while, so fair thou art," was one that I found strangely compelling and never forgot.

The second thread arose much later, after I had lived in Lexington, Kentucky, for some time. In 1980 the University of Houston offered me a teaching post. It would be a good job, but I didn't want to leave Kentucky. Then, one magical evening, my dog and I walked out through the university farm behind our house. The dog ran through the high grass. Waves of firefly light rippled outward as far as the eye could see. It was a night of such perfect crystalline beauty as to melt one's heart. And, in that moment, I knew I would accept the Houston job.

That bit of instinctive illogic puzzled me for years. Only when I finally ran across the full context of the Goethe quotation did I understand at last what had happened on that surrealistically lovely night in Kentucky. Faust had uttered his line while he was negotiating with the Devil. Faust, you see, did not strike a *bargain* with the Devil, as many of us suppose. Rather, he made a *bet*. Faust bet Satan that he could never be lured into settling down with any earthly pleasure—that his spirit would remain restless. The Devil agreed to the bet, and Faust rearticulated its terms in these words:

> Werd' ich zum Augenblicke sagen:
> Verweile doch! du bist so schön!
> Dann magst du mich in Fesseln schlagen,
> Dann will ich gern zu Grunde gehn . . .

That text has been translated many times and in many ways. I suggest this:

> When I say to the moment flying:
> "Linger a while; thou art so fair!"
> Then bind me in thy bonds undying,
> And my final ruin I will bear!

Eventually, Margarete becomes the moment to which Faust *does* say, "Linger a while, thou art so fair." His bet is lost, and Satan appears to have him.

Most English scholars accept Johann Wolfgang von Goethe as a Romantic poet, and Faust's cry was certainly a primary Romantic sentiment. A driving restlessness is the mainspring of the creative person. While they still bargain, Faust hurls his challenge at Satan: "When did the likes of you ever understand a human soul in its supreme endeavor?"

Goethe spent 41 years, on and off, writing *Faust*—from 1790 to 1831. During those same years, Watt's steam engines marched out of England and transformed the world. A new generation of fast printing presses made Faustian knowledge available to all in cheap textbooks. We began shaping a new world in iron. All the while the Romantic poets told us that we had the intellectual power to shape nature. "I will not cease from mental fight," cried William Blake. So, at the very instant I faced overwhelming contentment that night in Kentucky, I knew on some visceral level that I had to turn my back on it.

What Goethe captured was the sentiment that built Europe and America during the nineteenth century—Romantic discontent with anything but a world being recreated over and over, first in the human heart, then in the world beyond it. Marshall Berman uses Goethe's Faustian vision of the human passion for development as the springboard for his book on the nineteenth-century explosion of "development."[6] The technological establishment really did become Faust raging at Satan—saying that it was not about to ask any rare and perfect moment to linger. Goethe saw it coming. Technology and science in the wake of the eighteenth-century industrial revolution were Faust's intensity and fervor, made manifest.[7]

Now, like a stone skipping over water, that sentiment was coming to rest again. This time it did not land in the heavy machinery and classical physics of the nineteenth century, but rather in the wild new abstractions of the twentieth. And the first of our three quintessential geniuses, Thomas Edison, entered the scene while the stone was still in the air.

Edison was self-educated and isolated by partial deafness. At 15, he found work as a telegraph operator. That, in turn, led him to study electricity. He sat down and read everything Michael Faraday had written. At the age of 21, while he was working for Western Union in Boston, he applied for a patent (his first) of an electric vote-recording machine.

Edison was able to set up his own company by the time he was 29. He located it in Menlo Park, New Jersey, where he had some seclusion but was still only an hour's train ride from either New York or Philadelphia. The year was 1876, and, for the next decade, Menlo Park became Edison's season in the sun. The world was forever changed by his work there. He filed his five-hundredth patent near the end of that period.

Menlo Park was a small operation. One building housed Edison's office and a library. The labs and shops were in a second two-story building. He surrounded himself with a coterie of very bright engineers, scientists, and technicians. Historian Thomas Hughes tells how Edison created a complex, delicate, and unique collaboration.[8]

Out of Menlo Park came improved telegraph and telephone systems, dynamos, a precursor to the fax machine, electric rail systems, and the photoelectric effect. Menlo Park gave us the final practical development of incandescent lighting systems. It yielded the phonograph. Sound reproduction was an idea almost without precedent—a nearly deaf man's profoundly original gift to the world.

Edison's light bulbs were not as original as his phonograph, but to make electric lighting work, Edison revealed another kind of inventive genius. He developed the whole support technology for lighting, a web of electric companies. In 1880 he put his first complete electric lighting system on board a ship. By 1882 he had built New York's Pearl Street Power Station (mentioned in chapter 1), and he was providing the city with direct-current (DC) electricity for both lighting and electric motors.

Thomas Alva Edison, 1847–1931, from the 1897 *Encyclopaedia Britannica.*

Then, in 1884, his wife died of scarlet fever. That and, I suppose, success itself spelled the end of the greatest gush of pure invention the world had ever seen. In 1886, Edison remarried and moved into a much larger laboratory in West Orange, New Jersey. He continued inventing for another 45 years. More than half his patents lay ahead of him.

What Edison accomplished after Menlo Park would have spelled fame by itself, yet it paled against those ten years in Menlo Park. After that, says Hughes, Edison lost his incisiveness and his dramatic rendition. He lost the chemistry of that cadre of geniuses and craftsmen. The sun set on a great moment.

It was near the end of this period, in 1884, that Nikola Tesla came to America. The contrast between Tesla and Edison shows why Edison so solidly belongs to an earlier era.

Tesla was a Serb, born in Croatia and educated at the Austrian Polytechnic Institute in Graz. He went on to Budapest to work in the local telegraph office. After three years there, he obtained a letter of introduction to Edison, and he set out for America. He arrived in New York in 1884, just as Edison was having trouble with the electric lighting system on another ship, the steamship *Oregon.* Edison immediately hired Tesla and sent him to fix the problem.

Tesla succeeded, but he lasted less than a year with Edison. Their philosophical differences were insuperable. Edison said that 90 percent of genius was knowing what would not work. Tesla called that kind of thinking an empirical dragnet. He complained, "If Edison had to find a needle in a haystack he would proceed with the diligence of a bee to examine straw after straw until he found [it]. I was a sorry witness of such doings, . . . a little theory . . . would have saved him ninety percent of his labor."[9] Not only did Tesla's working philosophy part company with Edison's; his work with alternating-current electrical power also proved to be a huge obstacle in the relationship, since Edison was committed to direct current.

Tesla had first conceived of AC back in Graz in 1875, while he was still a student. He had brashly suggested that electric motors would run better on alternating current, and his professor told him there was no way to make such a machine. Tesla finally solved the problem of how to make an alternating-current motor after he had gone to work in Budapest.

And here Goethe reappears in our story. It happened one evening, very much like that evening in Kentucky, or like the quiet twilight in which Goethe began his tale of Faust. Tesla walked through a park at sundown. The problem of alternating current rode in the back of his mind while he recited a stanza from the play. The words were those of the aging Faust, before Satan arrived in his laboratory. Having failed to uncover the secrets of nature, Faust mused about sunset and the end of life:

> The glow retreats, done in the day of toil;
> It yonder hastes, new fields of life exploring;
> Ah, that no wing can lift me from the soil,
> Upon its track to follow, follow soaring!

It was in the line about *following* that Tesla realized what he could do. He saw how to rearrange waves of leading and following magnetic fields within a generator to produce the alternating current that has served us ever since.

Edison's working mode could not have been more different from Tesla's. Edison had adopted the nineteenth-century European idea of a research-and-development laboratory and made a huge success of it. However, Tesla recognized what Edison did not: genius was about to move off into far subtler arenas than straightforward R&D. Tesla never achieved Edison's business success, but he set us off in many directions that would shape the new century. The subtlety of alternating current in particular shaped the entire twentieth-century power industry.

Another of Tesla's profoundly important contributions was the first conception of radio. In 1893, Tesla demonstrated a primitive radio system at a lecture in St. Louis and used it to transmit an electrical signal without wires. The next year, Sir Oliver Lodge created a similar system and used it to show Great Britain's Royal Society that he could send Morse code over a distance of 150 yards.

Two years later, Italian inventor Guglielmo Marconi began work on a system identical to Tesla's, and by 1901, he had sent a signal that was reportedly heard all the way across the Atlantic. Up to then, none of these systems had transmitted the human voice. Since they all sent telegraph signals, the technology was first called *wireless telegraphy.*

Tesla, as he appears in his own photo—the first ever taken by phosphorescent light. (From the 1895 *Century Magazine*)

Thus Tesla, fueled by the Romantic imagination, built in his mind while more practical folk molded his ideas into reality. George Westinghouse created an effective AC infrastructure; Marconi completed the initial concept of radio.

Tesla was a true twentieth century prototype, a person from whom we expected nothing less than intellectual upheaval.[10] Yet, brilliant as Tesla was, it was Einstein who became the quintessence of twentieth-century genius. Einstein wrought such profound changes in the very framework of reality that he soon became a near-mythological figure. He fed his own myths and then used those myths to shield himself against a public that sometimes seemed ready to consume him.

Einstein's family had run an electric-machinery factory. Machinery and invention were central to the lives of the Einsteins. Albert Einstein studied applied physics in Switzerland and then became a Swiss patent examiner. He plied that trade until he was 30. He was good at it.

We tend to brush that experience off. Yet Einstein himself never lost interest in machinery and invention. He was an expert witness in patent suits through his thirties and forties. When he was 50, he and physicist Leo Szilard developed a new refrigerator. Szilard was also grounded in engineering, and together he and Einstein held eight patents. The key to their system was the electromagnetic Einstein-Szilard pump. They finally sold the idea to Germany's General Electric Company. Einstein also worked with gyrocompasses, and for years a Dutch firm paid him royalties on the compass he patented in 1926. In 1935 he invented a new airplane gyrocompass.

With that in mind, we go back and read Einstein's 1905 paper on special relativity. We discover that it dances with references to the same electric machine elements he looked at so critically in his patent office. His mental armory was well-stocked with engineering tools, with spatial concepts, and with machine elements.[11]

Einstein was born into the new industrial world of the late nineteenth century, which was shaped by a visual, machine-based, sense of the human potential. He invented a new physics that finally left that world

behind, but he himself never forgot his debt to that world or to that vision. To understand Einstein—to follow his clean, radical thinking in so many areas—we have to go back to that Swiss patent office. We have to remember his family's factory. Einstein knew what many other physicists often do not like to admit. It is that science is only great when it embodies a component of invention. Science has to do more than just report on nature.

One person in particular helped to lead Einstein away from Edison's world and into the twentieth century—the German philosopher Ernst Mach. Mach looked at the way in which science is often presented as a pursuit that moves along either of two roads. One scientist observes and measures nature; another writes mathematical theories. Within that model, experiment and thought seem to be separate roads to learning. Mach walked the experimenter's road; but, he said, we can also experiment in our heads. We need not limit experimentation to nature or to the synthetic nature we create in a laboratory.

Mach asked us to look, for example, at Galileo's reasoning. Aristotle had suggested that heavy things should fall faster than light ones—that a stone should fall faster than a feather. Galileo said, in effect, "Let's attach a feather to a stone and drop it. Together, they weigh more than either the feather or the stone alone. If Aristotle was right, they'll fall faster together than either would fall alone. But who could believe the feather wouldn't slow the stone? Aristotle had to be wrong."

We call that a *thought experiment*.[12] Galileo did the experiment in his head, and Mach made an odd argument to justify his doing that. He said that intuition is accurate because evolution has shaped it. We shrink from the absurd. That's what Galileo was doing when he said a feather cannot speed the fall of a stone.

Einstein was Mach's disciple, and, in his efforts to explain relativity, he created a great theater of thought experiments. But relativity flies in the face of intuition. As Einstein talked about throwing balls and reading clocks on railroad trains and elevators, he reached conclusions that departed entirely from intuition.

Mach finally cracked under the weight of Einstein's admiration. Relativity theory was, he wrote, "growing more and more dogmatic." For Mach, a theory was only a framework for our measurements. Einstein figured that one first writes correct theories; then the data necessarily fall into place. In 1919, Einstein shrugged off an important telegram. Someone wired to report that rays of light had been seen to bend under the influence of gravity, just as relativity predicts. Einstein was unimpressed. If light hadn't bent, he said, then "I'd have been sorry for the dear Lord. The theory is correct!"

So Einstein took Mach's thought experiments and used them in ways Mach never intended. He used them to show us a world far more subtle and surprising than anything our intuition had ever prepared us for. Mach's ideas about thought experiments finally faded because the world is much stranger than our capacity to imagine it. Einstein eventually wrote of Mach that his weakness was thinking that theories arise from observational discovery and not from mental invention.

There, in a nutshell, was Einstein's ticket into the twentieth century. For in that remark, he laid claim to the new concept of genius. Today we would expect geniuses not merely to thread their ways through complex thickets but also to surprise us with hitherto unthinkable realities. After Einstein, the world could expect a tidal wave of conceptual upheaval.

That tidal wave is exactly what we got. We move next to the place where we might well have seen it all coming, the new wave of modern art.

5

Remington to *Modern*:
Finding the Core on the Fringe

T he driving forces of change reveal their essential and primitive form, not at the core of an intellectual movement, but on the fringe. Once ideas lose their raw edge and begin finding a center of refinement, the forces that formed them blur and become harder to identify.

In chapter 3, for example, we see that the public's almost slapstick response to the new X-rays was an avenue through which modern physics raised general expectations. Chapter 4 describes the way hopes for the raw commodity of ingenuity spiraled upward for a season, after 1900. The popularization of genius in America reflected a deep change in expectations.

The fringes can be especially revealing when we try to understand how radical new forms of modern art evolved around the turn of the century. The French impressionist painters had set in motion a huge revolution just after the American Civil War. Impressionism was a radically new way of looking at the world, one in which artists captured their fleeting, momentary impressions of the reality before them.

Did those artists still report the same world we had always seen? Perhaps not. Impressionism was a very early stage of upheaval in a world where external reality itself was shifting. Artists began inventing new vocabularies to reflect new realities long before *Modern* established itself, and they created new realities as they found those vocabularies. It was very much a chicken-and-egg process.

Beginning around 1886, a new French school of postimpressionism took root. The postimpressionists went much further in augmenting the limited evidence of our eyes with material of the mind. French postimpressionists, such as Gauguin and van Gogh, created a huge emotional impact by taking us further and further away from visual literalism.

It would be simplistic to say they were driving new social, visual, scientific, and technological realities. It is more accurate to see them as a remarkably accurate and early mirror of what was going on around them. They were part of a resonance phenomenon—a mutual synergy.

While postimpressionists worked in France, American artists tried all sorts of experimental forms (including postimpressionism), but those forms do not fit easily into tidy classifications.[1] So let us move once more to the fringes to gain a glimpse of American art in flux. And it is on the fringe that we find Frederic Remington.

Remington was the archetypical realist, a seemingly pure creature of nineteenth-century rugged individualism. He was not at all someone we associate with *Modern*. Yet he grew. He eventually became a most unlikely revolutionary. He started out in one century and finally, if only by the skin of his teeth, did what few such people were able to do. He made it into the twentieth.

Indeed, Remington had only the most tenuous place in the artistic canon during his lifetime. His importance emerges only as we see it in hindsight. Remington's evolution encapsulates the change that was taking place throughout the world of art. Because he is the quintessential nineteenth-century man, his name is synonymous with the American Wild West.

But we've already seen what a gloriously contradictory place the West was. While it embodied the epoch of American imperialism, it was also a great stewpot of radical and creative thinking. By the time the politically conservative Remington finally reached the twentieth century, he had been thoroughly cooked in that pot.

Remington's name is linked with another nineteenth-century archetype, Teddy Roosevelt, and he can better be understood in relation to Roosevelt. Roosevelt was born in 1858 and Remington in 1861. Roosevelt went to Harvard; Remington went to Yale. Both were from wealthy families, each went West in 1883, and each eventually took up ranching. Each was deeply affected by the experience. Each lived with one foot set in the conservative Eastern establishment of the late 1800s and the other foot loose—ready for movement and radical change.

These two young men first converged in three 1888 *Century Magazine* articles. The 29-year-old Roosevelt wrote them, the 26-year-old Remington provided 33 illustrations. Brimming with the theater of ranching and cowboys, the articles show and tell us what ranching and cattle roundups were really like.[2] In them, Remington and Roosevelt shape the story of the American West as it's been told ever since. Here are Tom Mix, Gary Cooper, and John Wayne all in their original incarnations.

Remington had dropped out of college in 1881 and visited Montana. He published his first illustration in *Harper's* the next year. Then, in 1883, he bought a sheep ranch in Kansas. He worked at ranching and after that tried his hand at running a saloon and a hardware store. But meanwhile he carried sketchbooks and honed his abilities as an artist.

He soon found that he was far more interested in sketching the incredible world around him than he was in tending sheep.[3] He returned to New York in 1885, and he worked from sketches, photos, and memory to recreate what he had seen. He made another trip to Montana, Wyoming, the Dakotas, and Canada in 1887. He came back to show us his West in charcoal, watercolor and, eventually, oil and bronze.

The sickly and bookish son of privilege, Teddy Roosevelt had left New York for North Dakota, hoping to shoot a buffalo. He reached a region 125 miles due west of Bismarck and 200 miles north of Rapid City, South Dakota. When he saw the terrible hardships of the place and the harsh beauty of the scabrous land, he was enchanted.[4]

The year after Roosevelt arrived, his young wife and his mother both died—on the same day, in the same house—back in New York City. Roosevelt reacted by throwing himself into ranching over the next two years.

Another young adventurer had arrived in the barren western region of North Dakota about the same time. He was the handsome, 25-year-old French Marquis de Morès, who had married a wealthy American named Medora. The Northern Pacific Railroad had just laid rails across North America, and the Marquis formed a plan in that stark landscape. Instead of shipping live steers to Chicago to be butchered, he would use his wife's wealth to build a local packing plant. De Morès would butcher the beef first, then send it back in the new refrigerated railroad cars.

Roosevelt joined in with Marquis' ambitious plan and himself bought 450 head of cattle. Their money built a town, which the Marquis named after his wife. The new town, Medora, became the prototypical lawless

A Dispute Over a Brand, by Frederic Remington.

Line Riding in Winter, by Frederic Remington.

western boomtown. When the Marquis shot a man who was giving him trouble, he got away with it. Roosevelt called Medora a place where pleasure and vice were synonymous. The place buoyed his spirits once again.

As the Marquis began shipping meat to Chicago, he built a 26-room château and hired 20 servants for his wife. Roosevelt was quite taken by her, but he thought less of him. The Marquis eventually suspected that Roosevelt had undercut him in a legal matter and wrote to ask where Roosevelt stood. Since the Marquis was a notorious duelist, Roosevelt took the letter as a challenge. He wrote back to say that if the Marquis wanted a duel, he could have it. Lucky for him, the Marquis let the matter drop.

In 1886 the upward spiral of success came to a halt. The Marquis had expanded into projects that finally outreached themselves. First, his business empire caved in. Then the Dakotas suffered a terrible winter that killed three steers out of every four. Roosevelt returned to New York $24,000 poorer, but tanned, toughened, and self-confident. He eventually went on to form the Rough Riders, and he later became America's most colorful president. He created the National Park system, and he won the Nobel Peace Prize.

The Marquis did not fare as well. He went back to France, entered politics, and preached rabid anti-Semitism. Tuareg tribesmen finally murdered him in North Africa, where he was trying to join the French and Arabs in a holy war against the Jews and the English. The château is still there in Medora, but the cattle are gone, and Medora is home only to employees of the Theodore Roosevelt Memorial National Park.

A Row in a Cattle Town, by Frederic Remington.

Cattle Drifting Before the Storm, by Frederic Remington.

Soon after he returned to New York, Roosevelt saw some of Remington's sketches. Remington had perfectly caught the flavor of Western life as Roosevelt had seen it. So Roosevelt wrote those *Century Magazine* articles, and Remington illustrated them.

Of course cowboys don't do self-pity, and Roosevelt was now a cowboy. He wrote nothing about the fact that he had just been cleaned out. Instead, he rhapsodized about riding, roping, and six-guns. His cowboys had leather pants, yellow kerchiefs, spurs, and lassos. They boiled their coffee beans over open fires, and they ate from a chuck wagon.

The whole story is all there, firsthand, just as we've read about it for more than a century since. The articles were the work of two young men, neither of whom had finished honing himself. Roosevelt glibly proclaimed, ". . . a rancher's life is certainly a very pleasant one, albeit generally varied with plenty of hardship and anxiety." We see none of the human pathos within those hardships.

Roosevelt did not let his readers forget that he was really a New Yorker who had learned the ropes. He told of going out to search for a lost horse and being caught by a blizzard. He holed up in a hut with a cowboy from Texas. Roosevelt pulled out a small volume of *Hamlet* and read to the cowboy as they waited out the storm. Afterward the cowboy said, "old Shakspere saveyed human natur' some," and Roosevelt was greatly pleased with himself.

The introspective Henry Adams was a friend, if a critical one, to Roosevelt. Adams seems to have seen him as a nineteenth-century embodiment, out of place in the new century. In *The Education of Henry Adams*, he looks at Roosevelt, sighs, and says, ". . . all Roosevelt's friends know that his restless and combative energy was more than abnormal. Roosevelt, more than any other man living within the range of notoriety, showed the singular primitive quality that belongs to ultimate matter— the quality that medieval theology assigned to God—he was pure act."[5]

Still, it must be said of Roosevelt that he transmuted his primal affinity for action into a genuine love of the land, to which the National Park system bears testimony. He earned his place on Mount Rushmore. While he was surely one of our most intellectual presidents, his image of himself as a wild cowboy marked almost every moment of his remaining life.

Fifty-eight years after those *Century Magazine* articles came out, I moved to the small western logging town of Roseburg, Oregon, where I finished high school. One or two of the older students found summer work with the movies. They would fall off horses for five dollars a take. Next best to riding in the movies was riding in the rodeo. The movement of Western horse and rider still touched us all. We shared that deep-rooted vision of motion, masculinity, and rough-hewn grace that Remington had shown us first.

Remington's sketches, paintings, and wild cinematic bronzes moved John Ford to incorporate Remington tableaux into his Western movies. Ford openly copied not only Remington's composition, but his palette as well, to gain realism.[6]

Remington had grown up on a diet of military glory. As a green Civil War cavalry officer, his father, Major Seth Remington, had been named a hero after an ill-advised action near Centerville, Virginia, on June 27, 1863. When he rashly ordered his 90-man company to charge Confederate forces, it turned out that the enemy numbered 2,000 troops. Only he and 17 others came back alive.[7]

Remington was 15 when Custer's fiasco lay hold on the American imagination, and for years he tried to find that kind of glory for himself. After 1888, he made more trips. He chased after war and hungered to see battle firsthand. Meanwhile, he showed us action, always right at the point of impact—a rider being thrown from his horse, an Indian spear about to impale a pioneer.

While Remington reflected all the racism and imperialism that was America in 1900, his naiveté saved him. He just missed the slaughter at Wounded Knee; he just missed other combat. He finally found his live battle in the Spanish-American War. Roosevelt had recruited his Rough Riders in Texas and gone to fight in Cuba. Reporting for *Harper's Monthly*, Remington caught up with Roosevelt on June 30, 1898, at San Juan Hill. He finally had his chance to see real war.

But what Remington saw in Cuba left him nauseous. He looked at stacks of stripped and mangled corpses while guns made quiet popping sounds in the near distance. There was no glory there. Roosevelt came back a hero; Remington came back changed by the horror of it.

It took Cuba to bring Remington to introspection. Not long before his death at the still-young age of 48, his exhibits finally drew some of the critical praise he had always craved. He began painting quiet nocturnes—night scenes of a gentler West. He showed a West shrouded in winter snow—action suspended, humans and animals struggling to preserve their last calorie of warmth.

Remington finally formed a new artistic language in those calm and infinitely sad glimpses of the West by night. Images now emerged from a somber and wholly unexpected palette to show us the mysterious way in which people embraced hardship. These images are no longer the West of a boy's daydreams, but rather the artist's impression of a complex, imperfect, and terribly hard life.[8]

Their subject is no longer the "pure act" of his earlier work. What little action there is, is calm and muted—a trail-worn pack train threading through a somber mountain pass, coyotes lurking horses by starlight, a pair of elderly Indians chatting in the twilight. Remington takes us into the resignation of weary people around their campfires. No more of Roosevelt's nineteenth-century combative energy. If cowboys don't cry, they nevertheless need rest. They can sit still and come to grips with their own mortality.

Remington's action-packed West still defines the American sense of self, the Platonic ideal of a West that never quite was. What we remember of Remington is a West that objective eyes would *not* have seen. Remington's blind side let him present a West that was good and true. His early naiveté led him past the obvious folly, greed, and cruelty, and it allowed him to show us the beauty of our own dreams. But, in the end, it took an older and wiser Remington to form a new language. He created his own new form of American impressionism to tell us what it had all really meant.

The Herd at Night, by Frederic Remington. (This *Century Magazine* illustration reveals just a hint of the more introspective Remington who would emerge much later.)

And so the radical freedom of the American West, coupled with the visual feast presented by the land, fed artistic imaginations. That was no recipe for orthodoxy. Much of what the West produced was highly idiosyncratic, and it leaves you and me scrabbling to find the line between art and daydreams.

Indeed, the work of another artist was not only unaccepted during his lifetime, it was totally unknown. He was Charles Dellschau, an eccentric whose work really reflected the direction art was headed. Dellschau lived his whole life, which reached well into the twentieth century, in complete anonymity.

Dellschau emigrated from Prussia to Galveston, Texas, in 1850. He married a widow with a young daughter by the time the Civil War began, and he worked as a butcher in Fort Bend County, near Houston. Other than serving in the Confederate Army, he lived an unremarkable life. He had two more children of his own, and his stepdaughter married the noted saddle maker, Stelzig.

Then, in 1877, disaster struck the 47-year-old Dellschau. In rapid succession his wife, then his six-year-old son, died. Dellschau moved into Houston to work for Stelzig as a clerk. He stayed in Houston until 1923 and died at the age of 93. That would have been all there was to Dellschau's modest life, if it had not been for his secret hobby.

Somewhere along the way, maybe after he retired in 1900, he began drawing great airships. He filled scrapbook after scrapbook with incredible images. Lynne Adele, of the Huntington Art Gallery at the University of Texas, tells his story in the catalog of a traveling exhibit of self-taught Texas artists.[9] She also shows us his gorgeous, detailed and annotated, mixed-media images of heroic flying machines. They are Barnum and Bailey, Buck Rogers, and Jules Verne, all stirred together.

His pictures are mazes of exotic detail, circus-tent gasbags, bicycle wheels, belts and pulleys, crazily painted pods shaped remarkably like space shuttle booster rockets. Each fantastic vehicle has its own name: *Aero Mio, Aero Doobely, Aerocita,* and so forth. So much hope and feeling radiates from every picture!

Twelve of Dellschau's scrapbooks surfaced in a junkyard in 1967, 40 years after his death. His numbering system suggests that another 20 scrapbooks might have been lost. The surviving notebooks have found their way into art museums, and people have gradually been deciphering the coded writings he left with the pictures. As they've done so, a strange story has emerged.

Dellschau had belonged to the a group that called itself the *Sonora Aero Club.* It was a secret society that was formed in the California gold rush region around 1850. One of its members was supposed to have known how to distill a green crystal called *Supe* from coal. If one added water to *Supe,* it supposedly generated a gas that negated gravity. Naturally, when that Aero Club member died, the recipe for making *Supe* died with him.

Flight was on America's mind in the years just before the Wright brothers' first powered flight in 1903. Like many others, Dellschau was acutely aware that the ancient, deep-seated, and seemingly atavistic craving of our species was finally within our grasp. And, as we see in chapters 10 and 11, the wild, undisciplined American West was once more ahead of the game.

Several visionary airships actually *were* built and flown in California during the years when Dellschau was making his drawings. Perhaps those machines account for at least some of the sightings that people tried to attribute to Dellschau's mysterious Supe.

The West had more than its share of inventors, some drunk with dreams, others cold sober in their purpose, who were certain in their bones that they could fulfill the ancient craving. Dellschau's imaginings are tame alongside much of the machinery that was actually levitated into the sky during the early twentieth century.

But the secret self of the quiet clerk Dellschau reveals itself in much more than just another vision of flight. For Dellschau created a form of art that mirrored major movements in the European mainstream. It displayed the kind of expanded reality that postimpressionists had been exploring. It was embryonic surrealism. Delschau echoes ideas that were finding their way into the world of formal art, at the same moment.

That world, light years removed from Charles Dellschau, was also adopting hitherto unimagined forms and reflecting the new technologies. What had touched Dellschau and the famous artists alike was the pervasive and wide-ranging zeitgeist of *Modern*.

While the technologies (simultaneously new and primitive) of the American West drove Remington, and while the inchoate technologies of flight drove Dellschau, still another new technology was making a far more direct assault on art. That was photography. The camera has always been a paradoxical machine. It promised the ultimate visual literalism on the one hand, while it was actually the means by which we could see what our eyes had missed. The camera is not what it first seems to be.

Cameras were invented in two stages. The *camera obscura,* known for two thousand years, was a dark room or box that admitted an image through a pinhole or a lens in one wall. That light cast an image upon the opposing wall, where an artist could trace it. It was, in essence, a camera without film.

The second stage involved *recording* the camera's picture. The work of Joseph Niépce and Louis Daguerre finally resulted in daguerreotype photographs by the 1830s. The first exposures were slow. The pictures made by Daguerre and others tended to mimic the art of portraiture and of carefully composed still lifes.[10]

As better chemistry reduced exposure times, photographers became better equipped to catch the world on the fly. They did just what the impressionists were trying to do—to catch the image out of the corner of an eye, to preserve the fleeting moment. But photographers were attracted more rapidly to themes of the new industrial world than, say, painters were. They also turned to documentary reporting. They recorded the joining of the transcontinental railway on the one hand, and rotting corpses of the Civil War dead on the other.

From the beginning, photography became a battleground between external reality and inner vision. The evolving gender wars of the late nineteenth century provided a wild card in that struggle, since external reality and inner vision still had very different textures for women than for men. One nineteenth-century reality was that few women could lay their hands on this exotic new technology—too few to define a clear female perspective on that struggle.[11]

But two women did a great deal to reveal the potential of the camera, just because their perspectives were still individual and not part of any corporate consensus. They were Frances Benjamin Johnston, born in 1864 in West Virginia[12], and Anne Nott Brigman, born three years later in Hawaii.[13] Johnston took up photography in 1889; Brigman in the 1890s.

Brigman's camera created eerie, pantheistic visions. Female nudes flow organically out of wind-whipped scrub pines. The viewer cannot tell where nature ends and the human begins. There's a terrible intensity to her pagan celebrations. She writes, "Trees at high altitudes are squat giants twisted and torn with the sweep of the prevailing winds. . . . One day during the gathering of a thunder storm when the air was hot and still and a strange yellow light was over everything, something happened almost too deep for me to be able to relate."

While Anne Brigman walked the road of visual metaphysics, Frances Johnston played the hard-bitten professional. Her camera recorded America—its industry, presidents, schools, national parks, and social programs. The title of her biography, *A Talent for Detail*, comes from something she had written in the *Ladies Home Journal*: "The woman [photographer] must have . . . common sense, . . . good taste, a quick eye, a talent for detail, and a genius for hard work."

Johnston's pictures were subtler than she made them sound, however. She *created* reality with a sure use of irony while, ostensibly, she was merely *recording* it. For example, a picture shows students in one of the Indian schools set up soon after the brutal defeat of Indian nations. They are debating the question of citizenship for Negroes of the South. In a pair of self-portraits, Johnston strikes a proper Victorian pose on one side, while on the other she slumps, drinking beer, smoking a cigarette, looking tawdry, and exposing her lower leg. As she documents her world, she violates the canons and mocks that world. She does, in fact, alter the realities in front of her lens.

Different as they are, Brigman and Johnston display the common thread of rebellion. They both clearly announced that an old order was ending. Soon after World War I, the artistic revolution and women's suffrage had both been won, at least insofar as any revolution is ever won. It is in that sense that, when Brigman and Johnston pointed their cameras at the world, the world itself changed.

The person who did the most to bring photography into the artistic mainstream during the lifetimes of Brigman and Johnston was Alfred Stieglitz. Stieglitz was completely aware of the problems that cameras presented in relation to other artistic media. In 1902 he emerged at the center of a movement in New York City called "The Photo-Secession," whose aim was "to advance photography as applied to pictorial expression."

In 1905, Stieglitz and The Photo-Secession opened a gallery at 291 Fifth Avenue. The *291*, as it was called, became a center and arbiter of photography as art.[14] Both Anne Brigman and Frances Johnston were part of The Photo-Secession. Although Stieglitz was a lukewarm supporter of women photographers, he very much liked Brigman's ethereal style. The more down-to-earth Johnston gradually fell away from the movement.

Stieglitz struggled tirelessly to give art photography a legitimate place, yet he was very clear on one point. It was that painting and photography were two distinct art forms. As modern art sought out new realities, he said, painting would go where photography could not follow.

As a corollary to that, he also felt very strongly that, once photographers begin altering their images by hand, the result is no longer photography. Stieglitz would certainly not have seen a digitally altered photograph as representing the work of a photographer, had he lived to see it. Indeed, he would not accept even the cropping of a picture to remove extraneous ideas.

Beginning in 1908, Stieglitz, now a major figure in the New York City art scene, began to assert his profound understanding of the way in which the world was moving. He first created exhibitions of both art and photography at the 291. Between 1908 and 1911 he and photographer Edward Steichen exhibited Matisse, Toulouse-Lautrec, Rodin, Cézanne, and Picasso along with numerous photographers.

The Terminal by Alfred Stieglitz, 1892, from the June 1913 issue of *Camera Work*.

Woman with Mandolin by Pablo Picasso. (Published by Alfred Stieglitz in the June 1913 issue of *Camera Work*.)

All of Stieglitz's efforts were building toward the most famous art show of them all, the Armory Art Show.[15] But when that show finally came to pass, Stieglitz's role was that of a consultant, not a prime mover. In 1911 a group of 16 young artists formed the Association of American Painters and Sculptors—the AAPS. By now it was clear that the powerful National Academy of Design had established itself as the dictator of American tastes. Indeed, the academy had been the primary obstacle to Remington's acceptance as a legitimate artist. Now the academy was trying to ignore the huge artistic revolution, and the AAPS was not going to let that happen.

The AAPS named J. Alden Weir, also a member of the academy, as its president. The newspapers then announced that the AAPS had declared war on the academy. Weir was furious. He resigned the presidency. He claimed he had no idea the AAPS was so radical. Viewpoints hardened overnight.

The AAPS's first order of business was to exhibit the new art. It set out to do, in one grand exhibit, what Stieglitz had been doing piecemeal. The goal was to lay it all out before the public. But the AAPS could find no place to mount such a show. Madison Square Garden cost too much. Everything else was too small. Then one member said, "Let's rent an armory." That was a stroke of both genius and irony.

By that time, the wild, fanciful armory, made in the style of a medieval castle, had become a staple of city life.[16] American cities had started erecting fortresslike armories after railway workers went on strike in Martinsburg, West Virginia, in 1877. Workers had formed labor unions after the Civil War to combat the widespread bad working conditions that accompanied rapid industrialization.

Rioting and bloodshed began occurring in all the big rail centers. Cities came to fear their own workers, and they came to fear the immigrants who made up a large part of the work force. One magazine offered this appalling comment after the 1877 strike: "If the club of the policeman, knocking out the brains of the rioter, will answer, then well and good; . . . [If not] then bullets and bayonets, canister and grape . . . the way to deal with a mob [is] to exterminate it."

And so the armory was literally the castle keep for the militia. Most were National Guard headquarters. A typical armory had a huge indoor drill field, stacks of guns and ammo, lush veterans' meeting rooms, slits to shoot through, a great thrusting watch tower, and balconies from which to pour boiling oil on rioters.

The old armories were the oddest symbols of conscious paranoia. Turn-of-the-century architects announced that a building should proclaim its purpose. A church should be welcoming, a jail should be oppressive. And, of course, an armory should be "strongly suggestive of a fortress."

Those buildings proved to be far better as symbols than they were at fulfilling any function. People began to question their utility just before World War I. For one thing, a fire in an armory stuffed with incendiary goods could set it off like a Roman candle. But, more important, we soon saw that when class warfare did erupt it was a great deal more complex than the work of medieval peasants. And that is what artists were trying to tell us.

Still, those strange buildings persisted. They were familiar sights when I grew up in the Twin Cities during the 1930s. When I had the opportunity to go back to lecture on the University of Minnesota campus in 2000, I found one I had forgotten during the intervening years. The campus ROTC headquarters is located in one of the best-preserved of those old armory buildings, one built in 1896. I climbed the tower, strode the battlements, and became a character in some child's book about King Arthur. The building had all the old loopholes, turrets, and fairy-tale proportions.

The armory that the AAPS rented was the 69th Regimental Headquarters of the New York National Guard. For $5,000, the artists gained access to its vast floor space. Into that space poured work of the postimpressionists and the first moderns: Van Gogh,

The University of Minnesota Armory, now serving as the ROTC building. (Photo by John Lienhard)

Braque, Cassatt, Seurat, Munch, Matisse, Hopper, Picasso, Bellows. The great European and American art of that age reached New York, and the exhibit opened to four thousand visitors on February 17, 1913. Newspapers made news of it any way they could. They ridiculed the art, but no matter. The public had seen it, and they knew that change was upon them.

Just as the exhibit closed, 1,200 striking workers marched into New York from Paterson, New Jersey. The same intelligentsia who had backed the Armory Show had organized those workers. The revolutionary connection was quite explicit. The world was going to be changed. There was no shaking off this new vision of the human condition.

The AAPS did not survive the exhibit, but it didn't have to. The intensity of this gathering, the counterpoint, and the revolution being housed in that counterrevolutionary armory all conspired to give us new eyes. They changed our thinking in ways that we are still trying to understand almost a century later. Our new science-fed machines, and the social upheaval they brought with them, were shadows that wove themselves about every brushstroke of the new artistic forms.

Artists abstracted the new machinery and its motions. In so doing, they explained technology in terms that could not otherwise have been under-

stood. Duchamp's *Nude Descending a Staircase* was the star of the show. A buyer managed to pick it up for a scant $324, just after it had changed history.

A classical, if sinister, example of the synergy between art and the new machinery occurred soon after April 26, 1937. The Nazis, in support of Franco, had turned their Stuka bombers loose on the little Spanish town of Guernica to see just what they could accomplish. When the smoke cleared, they had killed or wounded 1,600 civilians. Spanish artist Pablo Picasso seethed with outrage, and he had just been given a perfect outlet for his anger. He had been commissioned to provide a mural to the Spanish Pavilion at the New York World's Fair. The fair was to be a kind of summary statement

Nude Descending a Staircase
by Marcel Duchamp.
(From *The Century Magazine,*
April 1914, p. 822)

of *Modern.* Into the Spanish Pavilion, Picasso took a vision of reality just as disorienting as quantum mechanics or relativity.[17]

The cartoon figures in his painting *Guernica* unleash his rage over the bombings in an artistic language so foreign that it hits us in the stomach only after we've turned away to look for more serious art. A German guide called it something any four-year-old could have painted. The Soviets didn't like it because good propaganda art was supposed to be more explicit.

But art is far more effective than explicit propaganda because it works down in the pit of our stomachs and connects with the primitive and childlike forces within us. *Guernica* worked that way on me when I finally saw it in Madrid's Prado Museum. I found the actual viewing not nearly as powerful as the accumulated recollection of it, later on.

Modern art achieved much of its power by the process of fusing a child's recollection with the temporal adult mind. That came home to me, very strongly, after a friend stopped by my office one afternoon,

The Eameses' children's construction cards.
(Provided by Margaret Culbertson; photo by John Lienhard)

several years ago. She delivered a three-inch stack of tough 7-by-11-inch cards with slit sides. Each card had a different design: there were spirals, gems, dodecahedrons, snowflakes. They were her childhood toys. "See," she said, "you fit them together and make things."

That evening, my wife and I built a castle from them while our cats, curious and worried, stalked about the work. Those cards were one of a thousand designs of all kinds of commonplace items created by two of the later modern artists, Charles and Ray Eames. Their collaboration began when the architect Charles Eames and the artist Ray Kaiser married in 1941, six months before Pearl Harbor.[18]

The Eameses had already invented something they called their *Kazam machine,* and they began using it to mold heated plywood by pressing it against a plaster mold. First they made molded plywood litters and splints for wounded soldiers. Next they made plywood airplane parts. As the war ground down, they turned to chairs. First children's chairs, then all-plywood dining chairs. Next they added chrome-plated steel legs. After that, the catalog of their work is a catalog of furniture you and I grew up with.

You've seen those chairs so often that you *don't* see them anymore. Seats of molded plywood are fitted with clean chrome-plated legs. Some-

(Photo by John Lienhard)

times black leather padding is added to the plywood. From chairs, the Eameses went on to create a whole range of office furniture. Then they molded fiberglass. They made those familiar black plastic and chrome assembly hall chairs that nest together so they can be stacked, 12 high. In another form, their chairs came to define the waiting areas in modern air terminals.

The hallmark of their designs was clean, childlike simplicity. The theme of *play* runs in all their work. In 1951 they made something they simply called *The Toy*—a set of large triangles that children could use to build playhouses.

They invented those mind-stretching building cards in 1952. They invented a coloring toy and a line of tops. They made movies, and they designed exhibit halls. For 40 years Charles and Ray formed the perfect symbiosis. "He was the energy, she the taste," a one-time Eames associate told me. For 40 years, their furniture carved its place in our lives.

The morning after we built that castle of cards my wife and I discovered that the cat had been in it. She had bent it into a new form, one that had not occurred to us the night before. Like all the Eameses' work, the simplicity of those cards hid possibility. They harbored the twin design virtues of simplicity and surprise.

Those virtues define another modern artist whose work has the same functional center as Ray and Charles Eames's did. He was Alexander Calder, an artist I cannot think of without summoning up the joyous orange crab that identifies our Houston Museum of Fine Art. It is a giant steel stabile crouching beside the front door of the original museum building.

Calder's wonderful abstract structures include countless mobiles of balanced steel plates and the so-called stabiles, which sit solidly upon the ground. I especially like the way he evokes animals with segmented slabs of steel. Whether or not you know Calder's work, you have certainly seen countless imitations of the style he created. He is one of the great modern artists whose work (like that of the Eameses) has settled into the background imagery of everyday life.

Calder was born in 1898, and art historian Joan Marter tells about the artistic silver spoon in his mouth.[19] Calder's father and grandfather were both noted sculptors. His father oversaw sculpture for the San Francisco Panama-Pacific Exposition in 1915. When it was clear that Calder had a talent for mechanics, his parents built a workshop for him.

Calder grew up making things in his basement and hanging out in his father's studio. After attending high school in San Francisco, he chose to study mechanical engineering at Stevens Institute of Technology. His parents were pleased. They did not want him taking up art. After he graduated in 1919, it became clear that he had no patience with daily work in an engineering company. But the education—the knowledge of kinematics, materials, and manufacturing processes—stayed with him.

He worked at this and that until an epiphany changed his life. He was serving as a fireman on a passenger ship. One morning, at dawn, he lay upon a coil of rope watching the sun come up, fiery red, with the moon showing silver against the last of the night sky. "It left me with a lasting sensation of the solar system," he wrote. After that, he went to art school in New York.

Calder the artist committed himself to sculpture, and Calder the engineer found his focus. His mobiles are studies of balance and kinetics. Every piece is an exercise in materials science. His stabiles are sophisticated steel construction. Through the 1930s he honed his skills and forged his artistic maturity. He works became everything good design must be: economical, durable, and superbly evocative. Calder wrote: "How does art come into being? Out of volumes, motion, space carved out within the surrounding space. . . . Out of vectors . . . motion, velocity, acceleration, . . . momentum. . . . Thus [the elements of art] reveal not only isolated moments, but a physical law [relating] the elements of life."

Calder returned to engineering through art. In that I hear echoes of T. S. Eliot's admonition, in the poem "Little Gidding," that our explorations inevitably bring us back to where we began. The value of our journey is that it finally gives us the eyes to see our place of origin for the first time.[20]

The art of the early twentieth century did that as well. It took us back to a childhood that had begun in the Victorian world. But first it had to fragment, disperse, and reassemble that world. Historian Stephen Kern rightly calls the period from 1880 to 1918 *The Culture of Time and Space* in his exhaustive treatise on the role of art in bringing *Modern* about.[21] For this was an epoch in which finding a way back home required that people rearrange their vision of those dimensions.

Thus, the year before he died, Frederic Remington wrote something that revealed, quite poignantly, how he had arrived where he once started out, and now had come to know the place for the first time. "I sometimes feel that I am trying to do the impossible in my pictures in not having a chance to work direct but as there are not people such as I paint [it's] 'studio' or nothing."[22] Still, Remington had miraculously made the jump and landed in the twentieth century. He managed to get there just as surely as his grudging contemporary Henry Adams had.

Adams knew that the art world needed a new place to stand. He had thought about the Virgin of Chartres while he contemplated dynamos in the Hall of Machines, and he also wondered about the female principle in late-nineteenth-century art. Speaking of himself in the third person, in the *Education of Henry Adams*, he said this about the character of American art at the century's end:

> Adams began to ponder, asking himself whether he knew of any American artist who had ever insisted on the power of sex, as every classic had always done; but he could think only of Walt Whitman; Bret Harte, as far as the magazines would let him venture; and one or two painters, for the flesh-tones. All the rest had used sex for sentiment, never for force; to them, Eve was a tender flower, and Herodias an unfeminine horror. American art, like the American language and American education, was as far as possible sexless. Society regarded this victory over sex as its greatest triumph, and [Adams] readily admitted it, since the moral issue, for the moment, did not concern one who was studying the relations of unmoral force. He cared nothing for the sex of the dynamo until he could measure its energy.

Behind that remark lay Adam's awareness of the pure act of Teddy Roosevelt and of the attempted demystification of the dynamo. By now he had certainly seen Remington's West as well, that purely masculine land that was no more.

But the catch is that Remington had, at last, become an artist—no longer an illustrator. At the end, he plumbed his mind for *impressions* of what had been. Snow and night gathered in around his pictures as new artists moved in to fragment the old world, as scientists sought out the micro-

The Gossips by Frederic Remington, *Scribner's Magazine*, 1910.[23]

cosm within the atom, and as engineers began building skyscrapers in Laramie, Houston, and Omaha.

After 1900, the sun still came up each day. Trees were still trees and flowers were flowers. Childhood was still present in our recollections, still shaping us as it has shaped every generation. Yet the old realities were now totally new, just as we ourselves had turned into beings that would have been alien to America's inhabitants in 1880. We had become modern—Remington by an act of will, Roosevelt as a force to be transmuted by others, Adams agonizingly conscious of the transition, Calder by riding in on the crest of the wave . . . We had all *become* modern by *creating Modern*, without halfway understanding what we had really done.

6

Fires and the High-Rise Phoenix

T he summer of 1871 had been one of driest on record in the central states. Rainfall had been a scant 28 percent of normal. By autumn, Chicago's wooden buildings were dry tinder, but so too were the forests of northern Wisconsin and Michigan. Sparks hit the tinder simultaneously in two places at once, around nine o'clock on Sunday evening, October 8.

That was when Mrs. O'Leary's cow supposedly kicked over a lantern and started the Great Chicago Fire. But we need to look more closely at the conventional history of the fire. In the first place, hindsight has finally given Mrs. O'Leary and/or her cow a tentative pardon. Second, the Chicago fire was not nearly as large as the other terrible fire that took place simultaneously, 230 miles further north.

Just before nine o'clock, the citizens of Peshtigo, Wisconsin, were anxiously watching a copper-colored night sky. All day, smoke and stillness had hung over Peshtigo. Now a peculiar cyclonic wind picked up, and it was driving nearby forest fires into Peshtigo faster than anyone would have thought possible.

Peshtigo was a young, burgeoning logging center with a population well over two thousand people. It lay seven miles inland from the west shore of Green Bay, next to Michigan's Upper Peninsula. The citizens had been fighting forest and grass fires in the area for several weeks, and they seemed to be keeping them at bay.[1]

Then the wind shifted in such a way as to drive a firestorm of epic proportions into Peshtigo itself. It came upon the town with appalling speed. People could only run for the Peshtigo River. Many died of asphyxiation. Others ignited like Roman candles as they ran. Eyewitness accounts tell of children flashing into flame while their mothers tried to hurry them into the water. One man reportedly went mad when he found that the burning wife whom he had dragged to the river was actually a total stranger.

The death toll in Peshtigo was eight hundred, but the fire touched many other towns in the area. In all, some 1,200 people died. Measured in human life, the Peshtigo fire was five times worse than the Chicago fire. The *New York Times* started reporting the Chicago fire on October 9, the day after it occurred, and Chicago remained front-page news for the next two weeks.[2] The paper counted the dead and homeless. It reported property losses. It described relief efforts.

Only a few articles mentioned forest fires in Wisconsin and Michigan. One said that 150 people had lost their lives in the larger port of Green Bay. Another article mentioned more fires in Windsor, Canada, but these fragmentary reports were only a sidebar against the theater of the Chicago fire and the expanding legend of Mrs. O'Leary and her cow.

Peshtigo failed to reach the American public in 1871, in much the same way as Rwanda and Biafra failed to reach us, a century later. It lay too far out in the wilderness of the new country. Chicago, on the other hand, had enormous presence in American life. It was a rail center, a marketplace, and the largest gateway to the West.

In his treatise on nineteenth-century Chicago, William Cronon paints a stunning picture of the vast commerce flowing through it in 1871.[3] Over a billion board feet of lumber were being shipped from the city by rail, and Chicago packinghouses were providing about half the American supply of hog products. (Some people had begun calling Chicago *Porkopolis*.)

The wind had been up on that Sunday night in Chicago, just as it had been in Peshtigo, and the Chicago fire didn't burn itself out until Monday night. Chicago, whose population had recently grown by a factor of ten, was an overcrowded, wood-built, bone-dry city with a poor fire department. The fire destroyed over three square miles of city, killed between 250 and 300 people, destroyed 18,000 buildings, and left a hundred thousand homeless.[4] In doing so, it truly did interrupt the flow of nineteenth-century America. Small wonder that the Chicago fire took precedence over a far worse slaughter, outside the national line of sight.

My 1970 *Encyclopaedia Britannica* cautiously says that the exact cause of the fire was unknown. But my old 1897 *Britannica* flatly blames it on an overturned lamp. When I was young, the great urban legend told us, and we all believed, that the fire started when Mrs. O'Leary carelessly allowed the cow she was milking to kick over a lantern.

Richard Bayles, of the Chicago Title Insurance Company, went back into his company's files to find Mrs. O'Leary.[5] He learned that she lived in a small rear house off Dekoven Street. Behind her house was a barn where she kept five cows. She sold milk to the neighborhood. Bayles studied testimony from the hearing that was held after the fire.

A young man with a wooden leg named Pegleg Sullivan testified that he had been on the far side of Dekoven Street and had seen fire break out in the O'Leary barn—nothing about Mrs. O'Leary or cows kicking lanterns. Sullivan had a lot to say about that night. He told how he'd run across the street to the barn and released the animals.

The trouble with his story is that old insurance maps show a house and a high fence blocking Sullivan's view of the barn. Sullivan's testimony that he ran two hundred feet on a wooden leg, then fought his way through the fire in the barn, is also dubious. Sullivan testified that he went to the barn every evening to feed his mother's cow, which he kept in Patrick and Catherine O'Leary's barn. So Sullivan had been in the barn himself. Bayles thinks it more likely that Sullivan started the fire by some act of carelessness—by dropping his pipe, or maybe kicking over a lantern.

Mrs. O'Leary had meanwhile been home in bed when the fire started. However, the fire department ended the hearings quickly, before it could come out that they had been taking bribes. The city had been taking care of neighborhoods that could pay extra money under the table, at the expense of Mrs. O'Leary's working-class neighborhood. The kicked-lantern story grew as the tabloid press went after Mrs. O'Leary. She finally had to flee to Michigan. In those days, the Irish occupied the lowest rung on the social ladder, and she made a good target.

But the real reason Chicago burned was that no one was paying proper attention at a time when all the conditions for a terrible fire were ideal. Indeed, in defense of the fire department, they had spent the preceding 24 hours fighting another fire that consumed four city blocks. They were exhausted when the big Chicago fire began.

One condition that practically demanded an outbreak of fires in Chicago is directly related to the Peshtigo fire. Chicago, like Peshtigo, was formed of *wood*. Its streets were largely made of wood, its sidewalks were all wooden, and its bridges were made of wood. But primarily, by 1871, the lumber industry of Peshtigo was feeding a radically new kind of home building, one that had been created in the Chicago area in the 1830s specifically to make use of the prevalence of wood.

I talk in chapter 2 about the powerful influence of wood on American architecture. This architecture had, outwardly, been rather derivative, and it continued to be, into the 1830s. Colonial construction had some uniqueness of appearance, but its real uniqueness lay in the way builders adapted European styles to the availability of wood in America, as well as to a shortage of skilled craftsmen. One had to be a jack-of-all-trades to make use of all that wood.

European houses, built of masonry and cut stone, exacted a high cost in labor. When Europeans built of wood, they relied on heavy timbers, accurately fitted with complex dovetailed joints. Not only was that labor intensive, it required special expertise as well.[6] At first the colonists tried to imitate European methods of building with wood. They embellished buildings by copying stone with wood; they imitated the tops of columns, cornices, and mantels in wood. One could even find chimneys made from wood and then daubed with clay.

A present-day visitor to George Washington's home at Mt. Vernon discovers that the whole building is made of false stone. The outer siding is wood, carved to look like stone, then covered with sand embedded in gray paint. It creates a very effective illusion.

America did things with wood that would have been unthinkable in deforested Europe. And, since wood called for nails, Americans invented automated nail-making processes. From 1776 to 1842, nail production improved repeatedly until American-made nails cost less than the tax alone on imported European nails.

Naturally, it could be only a matter of time before the increased availability of wood and nails had to give birth to a new concept of small-building construction. Historians argue as to when and how this new concept came into being, but all vectors point to the Chicago area in the early 1830s.[7]

A false stone porch column at Mt. Vernon. Wood covered with gray sandy paint. (Photo by John Lienhard)

The conventional story (which may be largely accurate) begins with a man named Augustine Taylor. In 1833, Taylor built a new church, St. Mary's, in Fort Dearborn, just outside Chicago. He managed to erect a 36-by-24-foot building for only $400, using do-it-yourself carpenters.

Taylor eliminated much of the conventional construction of mortised beams and fittings. He replaced them with light two-by-fours and two-by-sixes set close together. He used studs and cross-members and held the whole thing together with nails—no joints. Regular carpenters swore it would blow away in a high wind. But it did not. So the first baptism at St. Mary's was already disturbed by sounds of hammering next door.

Taylor's idea caught on very quickly and this kind of fabrication came to be called "balloon construction." The houses were called "balloon frames." Perhaps that was because, compared with the old timber framing, it seemed as light and insubstantial as a balloon. However, the term can be explained in other ways as well. A whole family of similar words refers to similar structures. The *baleen* plates in a gray whale's mouth are an example. So too is the French slang word for a jail, *ballon.* That term refers to a grillwork of bars. The term *balloon frame* thus seems to have been less a term of contempt than a reasonably accurate descriptor.

Today, we have modified the original balloon frame slightly. The two-by-four stringers once rose all the way from the ground through the upper floors. Since the 1930s, or so, it has become customary to make a new platform of each story and to build the stories independently. The result is a form of the balloon frame called *platform construction.*

In either form, but particularly in the earlier one, the frame house behaves much like a woven basket. It is light, flexible, and tough. Stresses distribute themselves throughout the structure. Stories are told about tornadoes knocking such houses off their foundations and the houses rolling away unbroken, rather like tumbleweed. While that may seem a bit of a stretch, the sight of one of these houses riding on a flatbed truck is familiar to us all, and it could not be done with far more rigid European timber-frame structures.

The balloon frame swept America. It became the conventional form of house framing used today. It was a uniquely American and wonder-

Typical platform type of balloon frame construction. (Photo by John Lienhard)

fully flexible reply to wholly new circumstances. Timber-beam houses had been naturally squarish in shape. Now a great gallery of architectural possibility opened up, and people took advantage of it. They built bay windows, watchtowers, and gables. They created homes with steeples, cupolas, and porches. They added Victorian gingerbread.

Immigrants to the Midwest were very capable people in a land with few architects. They needed designs more than they needed carpenters. As a result, pattern books of house plans appeared in the 1840s, in the very near wake of the balloon frame. For the rest of the nineteenth century, house designs published in books and magazines became a dominant force in domestic American architecture.[8]

And what houses were born of this new wedding of design and function! The Addams Family's house, where Morticia, Gomez, and Lurch lived, had all the earmarks of a mail-order home. Surely Norman Bates's house in the movie *Psycho* was one. Fine, gaudy old houses like those would later become the stuff of bad dreams in a brave new utilitarian world. But the balloon-frame became, and remains, the means by which America houses itself.

Fine example of an old mail-order house being renovated in Calvert, Texas.
(Photo courtesy of Margaret Culbertson)

It was solidly established in Chicago during the generation before the fire. By 1871, the city was made of those rapidly built, closely packed, flammable, wooden structures (many of which had been built from published plans). Meanwhile, lumbering towns like Peshtigo fed America's (and Chicago's) voracious and expanding appetite for wood. Many of the downtown Chicago office buildings provided an illusion of being fireproof, for they were faced with iron (as discussed in chapter 2). But inside those iron façades was more wood.

After the fire, Chicago's burned-out center became a kind of architectural tabula rasa—a clean, blank surface on which to build. There was no question that the city would *be* rebuilt. The people, by and large, were still there. The railroad tracks remained. A great deal of what had made Chicago so important survived the fire.

Into the burned-out, empty space converged architects from all over the world—traditionalists who wanted to work in stone, others trained in the new mathematical arts of designing for steel. By the 1880s, their debates were being reshaped by two new developments. One (mentioned in chapter 2) was commercial structural steel, now available in accurate factory-made beams. The other was the rapidly evolving *elevator*. Together, these were about to yield a new kind of building; they were about to give rise to the *skyscraper*.

Without elevators, there could be little purpose in erecting any building higher than an average person could reasonably climb. That limited most structures to about four floors. By the time of the Chicago fire, elevators had existed in some form or other, all the way back to antiquity.[9] Roman ruins reveal evidence of human or animal-powered lifts in some buildings. Down through the centuries, arrangements of baskets on the ends of ropes have been used to raise people up to high perches of one sort or another.

Rudimentary powered elevators had appeared in the 1840s, still a generation before the Chicago fire. Factories had started using their central steam power systems to drive rope-and-pulley lifts. Those early systems were poorly suited to dwellings, or even to small office buildings. They were cumbersome and required having someone constantly on duty, stoking a boiler and lubricating valves.

An even more pernicious problem had to be solved before the public would give its general acceptance to any kind of elevator. People needed some guarantee that the cab would not fall like a stone if the power failed or a rope broke. Potential buyers looked at a small room riding on the end of a spindly cable and thought about the hanging sword of Damocles.

Elisha Otis, whose name we still read on elevators today, solved the safety problem in 1854. He invented a foolproof automatic braking system. If the elevator began to fall, two spring-loaded "dogs" engaged cogs that ran the length of the elevator guide rails. For a while after that, steam elevators became as popular as any such cumbersome system could become.

The steam elevator would never become a major transforming force in commercial architecture. However, one form of elevator did begin what the electric-motor-powered elevator would complete. Before cities had electric power, they had high-pressure public water systems. Such systems began to appear just after the Civil War, and the availability of pressurized water offered another source of power that could be used to lift elevators.

An early steam-powered elevator, undated except for the mid-nineteenth-century dress, from the *Wonder Book of Knowledge*, 1923.[10]

The first hydraulic elevator consisted of an elevator cab at the end of a two- or three-story plunger, or *ram*, rather like the ones still used to lift automobiles in service garages. The Vienna opera house installed such a system as early as 1844. But a serious problem with the hydraulic ram configuration is that it requires a hole in the ground below the elevator to accommodate the plunger. Since the hole had to reach as far into the ground as the maximum lift of the elevator, such systems were impractical beyond a few stories.

Consequently, a new variant of the hydraulic elevator appeared. It used a large, short-stroke hydraulic piston to drive a rope through a series of pulleys. A five-foot hydraulic piston stroke could thus lift an elevator 50 feet or more. The energy was still being supplied in a hydraulic cylinder by the central water main, but now a much greater force acted through a shorter distance. Between 1880 and the early 1900s, hydraulic elevators using pulley systems were made to serve buildings as high as 12 stories.

Even after the development of electric elevators began in the 1880s, the hydraulic elevator kept appearing in the high-rise houses that were springing up in crowded eastern American cities. In 1901 and 1902, 68 of them were installed in Boston alone. The Smithsonian Institution acquired a small hydraulic elevator in 1984. It had been installed in a five-story Boston house in 1902 and had run until it jammed 40 years later.

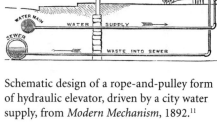

Schematic design of a rope-and-pulley form of hydraulic elevator, driven by a city water supply, from *Modern Mechanism*, 1892.[11]

When former Smithsonian curator Robert Vogel describes this lift (along with the whole forgotten technology of early domestic elevators) he makes the point that it is a technology most of us are not even aware of. The hydraulic elevator has long since been overtaken by cheaper electric systems. Still, it powerfully reflects the new impulse toward verticality that had enveloped American cities by 1900.[12]

All the new forces that gave (dare I say) rise to this new impulse converged upon the burnt-out streets of Chicago. Two Chicago architects, Louis Sullivan and John Wellborn Root, each worked quite consciously to create the new architecture of verticality. Sullivan wrote that the skyscraper, ". . . must be tall, every inch of it tall. The force and power of altitude must be in it, the glory and the pride of exaltation must be in it. It must be every inch a proud and soaring thing, rising in sheer exaltation that from bottom to top it is a unit without a single dissenting line."[13]

The new tall buildings began rising during the 1880s and early 1890s. The earliest ones were around seven stories high and built with heavy load-bearing masonry walls. But that was self-limiting; the higher the building, the thicker its walls had to be near the bottom. Root designed the ten-story Montauk Building, which was completed in 1882. It was supported by load-bearing masonry walls and served by a hydraulic elevator. Then Root pushed the idea to 16 stories in his Monadnock Building, finished in 1891. Its walls were six feet thick at the base of the building, and they had to rest upon a foundation that was 15 feet thick.

But the Monadnock Building, with its walls visibly widening outward toward the bottom, provides a fine example of the way old technologies outlast their replacements for a little while, for a new tall-building technology had already been created by the time the Monadnock Building was finished.

The secret to reaching really great heights was, from an engineering standpoint, kin to the Chicago balloon frame. It was the idea of creating a structural steel skeleton, and then hanging the walls and floors on that skeleton. Masonry will withstand only large compressive forces. Steel, on the other hand, is equally strong in tension and compression. Loads can be far more efficiently distributed through a steel frame than they can through stacked stone. As a result, much lighter buildings and far greater heights are attainable.

The first tall building with an internal structural steel skeleton was erected in 1885 (a date we shall encounter again). It was Chicago's nine-story Home Insurance Building, later extended to 11 stories. John Wellborn Root and his partner Daniel Burnham started another early steel-frame building in 1886 and finished it in 1891. It was the 13-story Tacoma Building. The race for height was on. Those buildings of the 1880s were the incunabulum of the new skyscrapers—the Empire State Building in its cradle.

An early Otis electric elevator, from *Modern Mechanism*, 1892.

While Chicago architects were learning how to erect tall buildings, German engineers were developing the electric elevator, and the American Elisha Otis came out with his own electric elevator scheme in 1889. Now practical buildings could rise upward in ways that had hitherto been unimaginable.[14]

The idea of very tall structures was, in itself, not new. Antiquity had given us the hundred-foot Colossus at Rhodes, the 440-foot Pharos lighthouse at Alexandria, and (before either of those) the 480-foot Great Pyramid of Egypt. Medieval Europe's Gothic cathedrals included the 460-foot spire of Strasburg Cathedral. While those structures were all masonry, none were segmented into multiple stories inside. They did not suffer the additional loading of one more stratum of heavy human activity for every ten or 12 feet of height.

As if to underscore the changing of the architectural guard, the 555-foot Washington Monument was finished in 1885. It edged out those ancient heroic structures to become the tallest masonry structure in the world, a status that it retains, even today.[15] The Washington Monument was dedicated the same year that Chicago's Home Insurance Building opened with its first steel skeleton. Together with the Monadnock building, the Washington Monument represented the last gasp of tall masonry construction.

Two other very important structures appeared just as the Home Insurance Building was finished, and they signaled just how far all this upward construction was about to go. The Statue of Liberty had a steel frame under its copper sheath, and the Eiffel Tower was a pure steel frame with the outer facing simply omitted.

Liberty set out by ship from Rouen, France, for New York City in May 1885. Three weeks later, *Scientific American* magazine celebrated her coming arrival on its cover. The picture is an artist's impression of Lady Liberty, and she looks far less solid and purposeful than the statue we know today.[16]

Scientific American grows more confident on the inner pages. There it offers schematic diagrams of the

Artist's conception of the not-yet-erected Statue of Liberty, from *Scientific American* magazine, 1885.

base and the iron skeleton. Gustave Eiffel had designed the inner skeleton shortly before he turned his attention to the Eiffel Tower.

The magazine tells how artist Auguste Frederic Bartholdi created the huge figure. First he sculpted her as a seven-foot model. Then he made an exact copy, roughly 35 feet tall. He cut the second model into sections and enlarged each section by a factor of four. He was then able to shape 3/32-inch copper plates to those pieces. Each plate was to be hung on the steel frame in such a way that it would impose no load on the other plates around it.

Scientific American had a great deal to say about the sculpture's size. The standard of comparison was the Colossus at Rhodes. Liberty was to stand 151 feet high—half again as high as the Colossus. Stationed on her pedestal, she would ultimately rise over three hundred feet above the waters of Upper Bay. Though *Scientific American* doesn't quote it, Emma Lazarus had already written her famous poem, "The New Colossus," which appears on the statue today:

> "Keep, ancient lands, your storied pomp!" cries she
> With silent lips. "Give me your tired, your poor,
> Your huddled masses yearning to breathe free,
> The wretched refuse of your teeming shore. . . ."

We might shrink from that "wretched refuse" line today, but the overall sentiment is still unique to America. Emma Lazarus names Liberty the *Mother of Exiles*.

Less known, however, are Lazarus' opening lines:

> Not like the brazen giant of Greek fame,
> With conquering limbs astride from land to land;

Never mind that the Colossus did not actually straddle the jetties at Rhodes; the point was clear enough: America meant to reclaim the grandeur of the ancient world and meant to do it by reaching upward. When Grover Cleveland dedicated Liberty a year later, her iron skeleton and been covered over, and that lovely lady, clad in cuprous oxide, changed the way in which we looked skyward.

Eiffel had meanwhile turned to the construction of his famous tower for the 1889 Paris Exhibition. This time, he used steel to far outreach any height ever achieved by any construction. The Eiffel Tower is 986 feet high, as compared to the 555-foot high Washington Monument. Eiffel now distributed stresses and made use of tensile loading within the tower's members, much the same way balloon frame construction did. The resulting weight difference between the Washington Monument and the Eiffel Tower is astonishing. The Washington Monument weighs 81 thousand tons and the Eiffel Tower, nearly twice as tall, weighs only *ten* thousand tons—only an eighth as much.

One way of visualizing the difference would be to suppose that we could scale each structure down to a height of one foot, reducing every part in precise proportion. Using a simple fact that weight is proportional to the cube of size, we have:

$$(\text{model weight/structure weight}) = (\text{model height/structure height})^3.$$

From this, we may determine the weight of such a model. If it were possible to construct perfect scale models, the Washington Monument model would weigh one pound (which is, in itself, light, since the monument is quite well built). But the one-foot Eiffel Tower model would weigh only *one-third of an ounce.*

This structural concept joined the new airplanes from the beginning. The earliest airplanes were wood, fastened with screws. (They were also close kin to the Chicago balloon frame.) Their strength was then increased when the frame was covered with a tightly-stretched glued-fabric skin.

Reconstruction of a World War I–vintage Curtiss Jenny at the Evergreen Flight Museum, McMinnville, Oregon, showing the delicacy of early airplane construction and its kinship to the Chicago balloon frame. (Photo by John Lienhard)

Many means of fabrication have been developed for use in airplanes, but the concept of stress distribution remains akin to balloon-frame and steel-frame construction. So let us make yet another scale model, this one of the Boeing 727.

The Boeing 727 has to withstand far more serious stress than the Eiffel Tower. It bears the huge forces of aerodynamics and of its own large engines. But those forces flow through its components and even through the stressed aluminum skin around it. Unlike the Eiffel Tower, it has wings and engines protruding on either side. Yet fully loaded with fuel and passengers, that model would weigh only two-thirds of an ounce. Stripped bare, it would weigh a scant eight or nine *grams.*

Eiffel thus made the next great leap forward in the search for height. Actually, he made not one, but *two* such leaps. Not only did he perfect the use of tension-loaded structural steel to attain great height, but he

1/3 oz. 1 lb. 2/3 oz.
Weights of exact scale models,
each 1 foot tall

also worked a miracle with the struggling new technology of elevator building. Ten years later Eiffel could have used one of the new electric elevators, but for now he had a problem, and the process of solving it created a great laboratory for elevator design.[17]

Carrying people almost a thousand feet upward was daunting enough, but Eiffel's more serious problem lay at the base of the tower. There, its four legs spread out, leaving a great open archway through the base. The arch is well over a hundred feet high and more than two hundred feet across. An elevator shaft in the middle of that lovely arcade would have been unthinkable. The ride to the top would have to begin within the curved, sloping legs that support the tower.

Hydraulic-drive cable elevators had never been used on such a scale, nor had they been used in the forbidding geometry of the tower's base. Eiffel hired a French firm to build a hydraulic elevator for the 550-foot run from the second landing to the top of the tower—one that began 380 feet above the ground. He had a much harder time finding someone to fit an elevator into the curved legs below.

Eiffel finally rejected a wild English screw-driven system and turned to the American, Otis. Otis was ready for him. He had been certain that no one else could do the job, so he had designed a system and then sat back to wait for Eiffel to come to him. Otis made two huge hydraulic cable elevators for two opposing legs. They were half elevator, half inclined railroad—very big and very complicated. They were also beautiful machines that drew crowds at the exhibition.

French nationalism demanded that French elevators be put into the other two legs. The French came up with a clumsy, but wildly inventive, chain-drive elevator that went only to the first platform. Later Eiffel muttered that its only merit was looking safe enough to satisfy bureaucrats. (They were the first elevators to be replaced after the exhibition.)

The Otis elevators were avant-garde technology, but before World War I, they also gave way to simpler systems. The surprise was the fate of the French hydraulic elevators in the upper part of the tower. They had once made the longest elevator run in the world, even though Eiffel had installed them with little fuss. They would have been replaced with an electric elevator because cold weather kept freezing the water in the hydraulic drive, but someone thought of adding antifreeze. After that, they kept running for a century.

Getting from the ground floor of a building to the floors above is a problem that stumbles upon a little-acknowledged aspect of invention. Our machines create metaphors, and, to succeed, new inventions have to follow old metaphors. The conventional elevator calls up what I call the "door into summer" idea. The story is told of a cat that goes from door to door on a cold winter's day looking for the one door that will open into summer—just as all doors did a few months before. The elevator likewise closes the door upon one world and opens it into another.

But, for millennia, the stairway has been our essential metaphor for getting ourselves upward. Far more obvious than the elevator is the idea of a special stairway that carries us upward—a stairway to the stars, the stairway to Heaven, a light at the top of the stairs. As my father grew old, he liked to quote a line in which T. S. Eliot grumbled about climbing the "penitential stair" at night. It is an interesting image. We've always wanted some magic stairway that would absolve us from the grueling penance of laboring up to the floor above.

Escalators seemed to fulfill the demands of the metaphor, but they were still rare when I was young. It was a heady adventure, in the 1930s, for a five-year-old to visit the one department store downtown that had one. Yet the idea actually reaches back to just after Elisha Otis's first patent. An escalator, with most of the features of modern ones, was patented as early as 1859.

Those early escalators, wed as they had to be to steam-power generation systems, produced too great a kludge for any practical use. But a second limitation to their widespread use was subtler. The capital expense required to build any escalator makes sense only when you want to move a large number of people. Because mid-nineteenth-century buildings topped out at around seven stories, a limited number of people used even the tallest building.

As buildings become far taller, their lower floors are dedicated to those services that most people in the building might use—the stores and restaurants in a tall hotel, for example, or the most heavily shopped departments of a department store. Today, escalators make excellent sense toward the bottom of such very high buildings.

So escalators lay fallow until the end of the nineteenth century. Then a spate of electric-powered moving stairways appeared.[18] The first was built in 1896, when Jesse Reno made a six-foot stairway that lifted great crowds of people onto the Coney Island boardwalk. Then, just four years later, the 1900 Paris Exhibition displayed four different kinds of escalators, including Reno's. Those escalators were catchy, but people still saw them as a fairground ride more than a functional necessity.

A Reno type of escalator, from the *Wonder Book of Knowledge*, 1923.

The attention that Reno got in Coney Island made him America's leading escalator designer. After Coney Island, his machines went into New York department stores and then into the Boston subways (another place where the high density of traffic justified their use). The Otis Elevator Company bought Reno out, yet the whole enterprise failed to reach very far beyond novelty. By 1920, no escalator company operated more than two hundred units.

Finally the Otis company sorted through the features of competing escalator systems and selected the best ones. Then the Otis company created what has been the standard escalator ever since—a machine that can carry eight thousand people an hour up a 30 percent slope. Before escalators became commonplace we had to recognize that they are not so much *labor* savers as they are *space* savers. Their real function is to keep people moving in crowded public places—like subway stations and the lower floors of certain high-rise buildings.

Still, the old Reno escalator was a solid piece of equipment, and many are still in use. You can spot them in the Boston and London subway systems by the characteristic structure of the wooden slats that make up their treads and risers.

Being swept up a moving set of stairs was high adventure for a child in the 1930s (and for the child's parents as well). The adventure sidetracked us all from realizing that the escalator was a subtle nexus of techniques that had more to do with completing the metaphor of going upward than it did with utility in high-density city life.

New York's early skyscraper, the Flatiron Building—in construction and completed, from the 1911 *Encyclopaedia Britannica*.

Thus the tall buildings rose, first in Chicago but then, very quickly, in New York. All the while, we struggled to accommodate a set of new technological metaphors. We also struggled with the issue of design. When the famous 20-story, 285-foot Flatiron Building was finished in New York in 1902, its lovely external stonework was rendered in Italian renaissance style. These new buildings did not immediately bring about a new architectural esthetic to match their disorienting verticality.

As late as 1911, the *Encyclopaedia Britannica* offered two distinct entries, one for steel construction and one for architecture. The article on architecture included one passing picture of the Flatiron Building and practically nothing else about the new vertical construction.

The section on steel construction went deeply into the technicalities of the new construction. It quoted the Chicago building code extensively and showed several photos of the Flatiron Building in successive stages of construction. None of that yet seemed to have any place within the old rubric of architecture. The first skyscrapers might well have been the first pure public expression of *Modern,* but they needed to rise still higher before we could recognize that expression for what it was.

Skyscrapers were now headed for a place that lay beyond utility—perhaps even beyond mere hubris. Not only did the smoke of Chicago and Peshtigo have to clear before America could see the awesome form of the Phoenix rising out of their ashes, but the form and shape of *Modern* had to establish a place in our lives as well: What the skyscraper was about to do would say much about the way we viewed ourselves as we stepped into the new century.

Once that happened we looked upward, expecting to see not just new tall *buildings* in the sky, but completely new *cities* as well.

7

The Titan City

Charles Jencks's book on skyscrapers offers a typology of tall buildings.[1] He suggests three essential forms: *skyprickers, skyscrapers,* and *skycities.* The Empire State Building, rising into a spire, is a classic example of a skypricker. When Jencks uses the term *skyscraper,* he refers to a tall, slablike building with a flat top. We see them all around us today, rendered in steel and glass. The Flatiron Building, with its isosceles cross-section and flat top, is a much older example of what Jencks calls a skyscraper.

The skycity is, for Jencks, the place where all this upward building is headed. His skycity is a cluster of interrelated tall buildings. It is an idea that began attracting architects, planners, and dreamers soon after the first appearance of tall steel-frame buildings. It was a *Modern* idea that went quite berserk, then seemingly died away. Yet, in another sense, it has crept back into our larger cities without our quite being aware of it.

One early theorist of the skycity, or what I think might better be called the *titan city,* was a most unexpected character. He was King Camp Gillette of Gillette razor blade fame. But, before he developed the safety razor, Gillette declared himself a radical socialist. In 1894 he published a book, entitled *The Human Drift,* in which he formulated his own titan city. It formed a part of his cry against capitalism, which he called "the most damnable system ever devised by man or devil." He also called individualism a *disease.*[2]

Gillette was born in 1855. His inventor/entrepreneur father and writer mother took him to Chicago when he was four. The family was wiped out financially by the Chicago fire, and Gillette was one part of the phoenix that rose from its ashes. He became a successful salesman with the Crown and Cork & Seal Company, all the while wanting to invent something great. William Painter, inventor of the Crown Cork (the first disposable cork-lined bottle cap), told Gillette, "Try to think of something like the Crown Cork, when once used it is thrown away, and the customer keeps coming back for more."[3]

Since it had obviously worked for his boss, Gillette systematically began searching for something to invent. It hit him the year after he published *The Human Drift*. One morning, while he was shaving, it occurred to him that he might make a disposable razor blade. At first he had no idea how to make good disposable blades. He struggled for years before he met a fine inventor named William Nickerson.

Gillette's idea was the perfect challenge for Nickerson, who finally saw that he could solve the problem by making an extremely sharp blade, thin and wide. The blade holder could then bend it into an accurate position and hold it there securely. Together, Gillette and Nickerson began making *safety razors* in 1903.

As to naming the new product, Nickerson's name was too suggestive of nicked skin, so they used Gillette's name and put his face on the razor blade wrapper. (Some 96 billion copies of Gillette's portrait had been circulated by 1948, according to one estimate.) After just ten years of producing blades, Gillette had become a multimillionaire, and he retired from active work in the company to grow dates in California.

That alone has the ring of a rather conventional success story, but it seems inconceivable against the other face of Gillette. The terrifying idealism of *The Human Drift* seems to have emerged from an utterly different person.

Gillette had imagined a Utopian socialistic world based on universal cooperation. One great company, with all people as shareholders, would do all the production. They would also do it far more efficiently than they did in a free-enterprise system. He wrote, "Selfishness would be unknown, and war would be a barbarism of the past."

The most bizarre feature of the Utopia he formulated was that all 60 million Americans would live in one great city, powered entirely by the Niagara River. Here, perhaps for the first time, the word *Metropolis* appears as a proper noun. Metropolis was to be a vast complex of apartment buildings with a hundred million rooms. The city was to occupy a 30-mile-wide strip running for 120 miles along the southern border of Lake Ontario. It would include Niagara Falls at the western tip of the lake and extend to a point just east of Rochester. Gillette wrote, "Under a perfect economical system of production and distribution, and a system combining the greatest elements of progress, there can be only one city on a continent, and possibly only one in the world."

And so, although Jencks makes no mention of it, the fully evolved concept of the skycity had already been right there in Gillette's mad vision. In a single step Gillette wanted to go all the way from individual buildings to a single great aerial city. He laid out a literal beehive in hexagonal city blocks. Each block held either an apartment or some sort of corporate utility building. The citizens of Metropolis would, for example, eat communally in vast dining halls.

The layout of city blocks in Gillette's Metropolis, from *The Human Drift*, 1894.

This was less than a decade after the new steel-frame architecture had come out of Chicago, but Gillette had already formed a surprisingly clear picture of how it worked. He used these words to describe the 25-story apartment buildings in which he meant to house America:

> These buildings are constructed upon the general plan of modern office buildings, such as are seen in cities like Chicago, New York, etc. It consists of a steel framework that is filled in between its network of beams and girders with fire brick, which constitutes floors and walls. These floors and walls are then covered by a facing of porcelain tile in every part of the building, both inside and out.

The weight of successive stories does not come upon the stories below, but are each separately supported upon independent steel beams and girders . . .

Despite his expressed scorn for individualism, Gillette realizes that humans will not put up with an ugly sameness. He offers the following relief: "Each and every building of 'Metropolis' would be a complete and distinct work of art in itself. Every color and every shade of color would be found in their ceramic treatment."

By 1910, Gillette had given up on Niagara Falls and obtained articles of incorporation from the Arizona Territory for his *World Corporation*. He also offered Teddy Roosevelt $1 million to serve as its president. Roosevelt declined. Later, Gillette turned to social reformer and writer Upton Sinclair, who gave Gillette a sympathetic ear and moral support for his ideas.

In an attempt to build backing for the idea, Sinclair set up a meeting between Gillette and Henry Ford. The meeting was a disaster. The two millionaires talked past each other, and they finished by simply shouting at each other in anger. Perhaps the outrageousness of Gillette's scheme was not the only thing that so offended Ford. The grandiosity of Ford's thinking in many ways matched Gillette's.

I suspect Ford realized that Gillette's Metropolis offered very little significant role for the automobile. The titan city was gaining ground in many people's thinking, and the automobile would prove to be a key agent in stemming the spread of the idea. (More on that in chapters 8 and 9.)

Sinclair also helped Gillette write his most cogent statement—a book called *The People's Corporation*.[4] Even that was naive at best. Stuart Chase wrote that Gillette's sincerity was deep and compelling, ". . . but his solution is quite untouched by the realities which guard the road to Utopia."[5]

In his last years, Gillette turned to the extraction of oil from shale—something that would become a major technical enterprise toward the end of the twentieth century. He obviously kept his inventive edge, as well as a dimension of real vision, to the very end. He was a wealthy industrialist who preached socialism and never did seem to understand that the

One of Gillette's typical Metropolis apartment structures, from *The Human Drift*, 1894.

simple elegance of his safety razor was not to be confused with the complexity of human affairs.

He was a man of his times, and his was not a time when we dealt reflectively upon our new technologies. Yet the capitalism of the early twentieth century (like that of the late eighteenth century before it) did have a strange dimension of full-speed-ahead social reform stirred in with it.

All the while, buildings rose like Kansas sunflowers. As early as 1913, New York's 57-story, 792-foot Woolworth Building had completely eclipsed anything Gillette had conceived for Metropolis.

By the time Gillette died in 1932, two truly enormous skyscrapers (in the conventional sense of the word) had been completed in New York City; and, together, they would define the word *skyscraper* for years to

Last (†) and first of the great New York skyscrapers: Left, the World Trade Center, completed in 1972. Right, the Woolworth Building, completed in 1913.
(Photo by John Lienhard)

Empire State building

Chrysler building

Midtown Manhattan skyline in the year 1999. (Photo by John Lienhard)

come. The Chrysler Building, completed in 1930, stood 1,046 feet and 77 stories high. Three years later, the 102-story, 1,250-foot Empire State building surpassed it. It would be 1972, long after *Modern* had become passé, that the World Trade Center would rise any higher than that.[6] After the Empire State Building was completed, people came to their senses for a while.

Well, most people did. In 1959, Frank Lloyd Wright created the preliminary design of the one-mile-tall Illinois Tower. Although the building offered many design problems, it might well have been made. That was the same year Wright died at the age of 92, and we're left to speculate as to whether the idea of such a building was, at that time, visionary or anachronistic.

Just before the Chrysler and Empire State Buildings were erected, the skycity had started to look like the true shape of the future. The image was widely agreed upon, and it was quite overwhelming. In the conventional view of it, huge buildings rose, layer upon layer. Multitiered highways flowed among the buildings on many levels. Very early in the twentieth century, pictures of such cities began appearing. They were copied, circulated, and made into postcards. The idea struck a nerve. The images showed flocks of the most primitive airships moving among the spires. In most cases, the ribbons of highway, high above the ground, were still traveled by trains, since the automobile had yet to catch up with the dream.

More often than not, these images were offered in a kind of pie-in-sky, daydreamy way.[7] But some were intended as savvy urban planning. An article in the July 26, 1913, *Scientific American* magazine has a cover picture that shows such a city with detailed cross-sectional views of its lower regions.[8]

We find no aeroplanes in this image. It shows, instead, a careful hierarchy of movement. Upper paths and bridges are meant entirely for pedestrian travel. They lie next to storefronts, far above the ground, rather like upper walkways in today's large indoor shopping malls. Roadways

Cover of *Scientific American* magazine, 1913.

below are for motorcars, omnibuses, and the occasional horse-drawn vehicle. Electric trolleys run in tunnels below the street, and in the layer below them are tramways for merchandise. The author likens the separated flow of pedestrians, vehicles, and materiel in a city to the handling of fluid flows. After all, one would never pass oil and water through the same conduit.

That kind of hydraulic segregation was an important part of the story of the city of the future. Whether fanciful or serious, all such pictures were careful to draw our attention upward. That *Scientific American* article was one of the few that acknowledged any world below. The bottom was normally out of sight. What we actually see in most such images is only the part that reaches *up*. The vision was Gothic, often including even the gargoyles and ornate towers of the old cathedrals. But now upper ribbons of highway replaced the old flying buttresses.

Wanamaker's department store made a formal assembly of this idea in October 1925.[9] The store put on an art exhibition and called it, "The Titan City, A Pictorial Prophesy of New York, 1926–2026." There, in painting after painting, was that Gothic upward reach of the city into the sky. It made theater of what we all expected the future to be. It was an expectation that would collapse only after the Tower-of-Babel impulse that drove it became more obvious.

Fritz Lang's movie *Metropolis* did much to expose that impulse. It came out of Germany the year after the Wanamaker's exhibit. Lang realized that such a city would necessarily *have* those lower depths that we did not see at Wanamaker's, or in other images. His Metropolis consists of a pampered upper class, living in luxury far above the ground, while an army of slave laborers toils in the hellish underground to make it all function.

So stratified is life in Lang's Metropolis that people in the regions above and below are unaware of each other's existence. A revolutionary leader of the underground emerges in the person of the laborer Maria, after she discovers the world above ground. Metropolis's "management" subverts the rebellion by replicating Maria as a robot who would lead the workers where management wanted them led. The robot, clothed first in brilliantly conceived armor plate, and later transformed to look like Maria's own flesh and blood, is one of the truly sinister ideas that the movies have permanently imprinted upon us.

After *Metropolis,* the idea of the titan city began to waver, and, for a season, heroic skyscraper building stopped. Of course, the Great Depression took its toll just as the automobile began making downtown workplaces accessible to suburbia. Yet, even before the Depression and the full impact of the automobile, *Metropolis* had taken the edge off a great expression of energy.

Indeed, movies keep reviving Lang's vision of hell. Central to *Metropolis* and all its progeny is that sinister, hidden underworld. During the past generation we've seen it in *Blade Runner, Batman, Dark City*, and *The Matrix.* You will certainly think of others, and you can bet that new ones will continue to emerge in the stories we tell to one another.

Compound themes of *artificiality* also wreathe our visions of the titan city—artificial food, artificial light, and finally, artificial life. We saw artificiality carried to one limit in *Matrix* and *Dark City,* where the worlds themselves became artificial. But the titan cities in *Blade Runner* and the movie *A.I.* were even more horrific when they picked up the central theme of *Metropolis*—that artificiality was poised to take over our humanity itself. The titan city generated its own incarnation of the Frankenstein story for the twentieth century.

For skyscrapers to survive into the future, they had to shed their Gothic clothing and find a style without *Metropolis*'s oppressiveness. I noted in chapter 6 that architects initially regarded skyscrapers as being the province of structural engineers. Ominous Gothic themes had settled in upon skyscrapers from the beginning, but they had done so by default. The Woolworth Building displayed Gothic flourishes. The titan city, which was never actually built in the twentieth century, began in Gothic style; and, when it resurfaced in movies at the end of the twentieth century, it invariably displayed those same dark qualities of Gothic menace. The new buildings clearly needed a new style.

The architects and dreamers who mounted the Wanamaker's exhibit found a visual vocabulary for the new architecture in the emerging style of *art deco*. Art deco was made to order for tall buildings. It had appeared in embryonic form as early as the 1913 Armory Art Show, and fully developed art deco was a creature of the 1920s. It was the functional answer to the far more ornate art nouveau of the preceding generation. Art historian Arie van de Lemme describes it in these words: "Where Art Nouveau had been heavy, complex and crowded, Art Deco was clean and pure. The lines of Art Deco did not swirl about like the centre of a whirlpool; if they curved, they were gradual and sweeping, following a fine arc; if they were straight, they were straight as a ruler."[10]

That, of course, sounds like a prescription for skyscraper design. Indeed, some of the early skyscrapers had already picked up on art deco themes, and, in 1933, the architects of the 850-foot Rockefeller Center took the idea to its limits. They not only used the vertical lines of art deco in the exterior design of the building, they also filled the public spaces with art deco bas-relief and surrounded the building with art deco statuary.[11]

Rockefeller Center was far less sinister than Gillette's Metropolis or Lang's, yet it was a direct outgrowth of the titan city vision. A few years earlier, the Empire State and Chrysler Buildings had also been very pure examples of art deco. The theme was less elaborate than it was in Rockefeller Center, but perhaps stronger for just that reason. The Empire State Building displays the lines of art deco with dignity and grace. It tells us what art deco can be at its best. But art deco should not be too dignified.

The glorious spire on the Chrysler Building, far more than the plain-vanilla, never-used, dirigible mooring mast on the Empire State Building, is the most powerful and pervasive image of art deco that we have. It touches millions of New Yorkers every day, even now. Architect William Van Alan designed that wonderfully dramatic structure with its magnificent spire; but behind it lay the motive force of Walter Chrysler.[12]

The Chrysler building.
(Photo by John Lienhard)

Most of us know Chrysler only as a last name. He was, in fact, a remarkable and flamboyant exemplar of *Modern*. His impulses, like Gillette's, were populist to the core, but without the sinister undertones. Chrysler had truly come up from the workshop, and this building was his joyous, self-expressive boast. He had made it to the top.

In 1907, Chrysler was a talented 32-year-old mechanic working for the Union Pacific Railway.[13] That year Chrysler visited the National Automobile Show in Chicago, where he fell in love with a $5,000 luxury automobile called the *Locomobile*. He went into hock to buy it. He couldn't yet drive, but he put the car in his garage and set about taking it apart and putting it back together.

Chrysler soon left the railway and went to work for the embryonic Buick Company. After that he bounced from one success to another until, as CEO of the Chrysler Corporation, he was selling more cars than Ford. The early automobile makers had once been in the carriage business. They were workers in wood. Chrysler had better instincts for mass-producing with metal.

Chrysler's wife tried to interest him in opera, but he loved showy musicals, and he kept company with Flo Ziegfeld. While art deco was extraordinarily popular in the 1920s and '30s, it was also considered gauche. Naturally, Chrysler loved the style. So it was that, in 1934, he gambled on the glitzy Chrysler Airflow, which combined art deco with the even newer style of *streamlining*.

Most cars still looked like somber boxes. The Airflow was shaped a little like the later Volkswagen Beetle. Back then, I didn't know the term Airflow. We all called it by its advertising nickname, "The Car of the Future." The Airflow broke ground in many ways. It was a well-engineered machine with a fine new suspension system. But I did not find it especially pretty when I was a child, and I still think it was a visually unbalanced design. The rest of the American public felt the same way. They wouldn't buy it.

We would play the game of counting Airflows when we took trips in the 1930s. We never saw many, but the cars that came after it copied and softened that design. Chrysler's Airflow did away with the old boxy shape, and it redefined car design just as surely as the spire on the Chrysler Building continues to announce the New York skyline.

Chrysler, by the way, never lost contact with metal, and he never lost contact with his public. When he saw a stalled car on the road, he would often stop, get out his toolbox and give aid. Then he would hand the surprised driver his card and suggest, "Next time, maybe you should buy a Chrysler."

Today, many of us still recognize Henry Ford's hard face, but who has any lingering sense of this garrulous, people-loving person? The name *Chrysler* does not evoke a face at all. Nor, for that matter, does it even evoke any particular automobile (even though his cars are still with us).

The image that enters our mind is the glitzy building that brought an epoch to its apogee—the very emblem of America reaching skyward too rapidly. This was the full expression of the idea of the skyscraper (what Jenkins would call a skypricker). After the Chrysler Building, where could one go but directly to the titan city?

And so thinking about Chrysler gives us a framework for going back to look at Gillette. The same populist themes surround both, but the dimension of human freedom somehow got lost in Gillette's thinking. His socialism and capitalism fused to regiment and divide his world—which (as Lang reminds us in *Metropolis*) is a persistent danger in Utopia.

H. G. Wells sounded precisely the same warning back in 1895, only one year after Gillette published *The Human Drift*. In *The Time Machine*, Wells's time traveler visits a distant future in the year 802,701. As Lang would do 31 years later, Wells also shows Gillette-like thinking being doomed to create a divided society.

The pastoral and peaceful Eloi live above ground; the ferocious Morlocks live below. Morlocks come to the surface now and then to capture and eat an Eloi. Then we discover that the Morlocks maintain the machinery below and do what is needed to serve and supply the totally dependent, peace-loving Eloi who are, it turns out, their cattle. Thirty-one years before Lang rang a new set of changes upon this theme of the modern underworld in film, Wells had already carried it to its logical conclusion. He too shifts our sympathies from the Eloi above to the Morlocks below.

Like Gillette, Wells went on to formulate alternate social orders. He traveled the globe trying to develop support for the idea of a single World State. He sought to create a single religion that was pure and undefiled.

The impulses of these two dreamers were remarkably similar. Yet that Gothic dungeon below, that home of the Morlocks, has always been the undoing of the titan city.

Five years later, only six years after Gillette published *The Human Drift*, Wells wrote another novel: *When the Sleeper Wakes*. This one is far less familiar than *The Time Machine*, but Wells uses it to address the concept of the titan city directly.[14] Wells sets his city two hundred years in the future, in 2099. Although the city is encased in glorious glass and steel, it is still Gothic. Wells writes, "A cliff of edifice hung above him . . . the opposite façade was grey and dim and broken by great archings, circular perforations, balconies, buttresses, turret projections, myriads of vast window, and an intricate scheme of architectural relief." This time, Wells is explicit in predicting the presence of a supporting underclass below all the apparent beauty. He describes it thus:

> Presently they left the way and descended by a lift and traversed a passage that sloped downward, and so came to a descending lift again. The appearance of things changed. Even the pretence of architectural ornament disappeared, the lights diminished in number and size, the architecture became more and more massive in proportion to the spaces as the factory quarters were reached. And in the dusty biscuit-making place of the potters, among the felspar mills in the furnace rooms of the metal workers, among the incandescent lakes of crude Eadhamite, the blue canvas clothing was on man, woman and child.
>
> Many of these great and dusty galleries were silent avenues of machinery, endless raked out ashen furnaces testified to the revolutionary dislocation, but wherever there was work it was being done by slow-moving workers in blue canvas. The only people not in blue canvas were the overlookers of the work-places and the orange-clad Labour Police.

Wells gave us Lang's Metropolis in words instead of pictures. And this passage alone clearly displays why the titan city was an excess of *Modern* that could not take root. Our innate fear of the titan city's lower depths festered from the beginning.

Yet the titan city is finally edging its way into the world of the twenty-first century. It is doing so quietly but relentlessly. Minneapolis now has an elaborate skein of upper-level skyways throughout its downtown. They provide protection from the winter cold, while they slowly mold Minneapolis into something very near to the city of the Wanamaker's exhibit.

For years Houston has also had a downtown web of interconnection, but that web has been located underground to protect people from the summer heat. Now we find, as we look upward, that Houston is taking advantage of public air conditioning, and building a system of interconnecting skyways *above* the streets, as well.

The concept of the city as interconnected elements, rather than as functions isolated in independent buildings, is an idea that is feeling its

An overhead connecting passage in downtown Houston. As such connectors become common in downtowns across America, the titan city emerges in a more utilitarian incarnation. (Photo by John Lienhard)

way into our urban fabric. The titan city emerges as we identify the loci of value that were embedded in it. Once we shed the vast hubris that took hold of the idea during the latter days of *Modern*, we go back to seek out those values.

By the time the Great Depression brought us all back down to earth, the Faustian dimensions of skyscraper building, and of titan city dreams, had become pretty clear. Perhaps the titan city idea would have died even earlier than it did, had art deco not given it a brief reprieve.

But the image of a whole city in one vast building kept its dark Gothic overtones, and its sinister underbelly, throughout the twentieth century. Perhaps the idea was doomed simply because we knew, on a gut level, that our kinship with the Morlocks was greater than our kinship with the Eloi. We gazed upward at the glistening tip of the Chrysler Building and knew that it was meant for someone else.

Besides, by the time Chrysler had erected his building, he was a major maker of automobiles, and automobiles directed our attention in exactly the opposite direction. The people of the city were now able, for the first time, to look *away* from the city center in the conduct of their everyday life. And it is to that matter that we turn next.

8

Automobile

The 41-story First National Bank Building in St. Paul, Minnesota, was, by far, the tallest thing in my childhood experience. By the turn of the twenty-first century, only two St. Paul buildings had topped it, and they by only a small margin. Finished the year I was born, the First National Bank Building dominated the city, and it has remained the first image I see when I hear the word *skyscraper*.

However, a new family ritual was already undercutting that image. When I was around five, my parents bought a patch of land east of town, just south of the Hudson Road to Wisconsin. It lay on the shore of small Wilmes Lake, next to a 350-acre truck and dairy farm. My father hired a carpenter to build a small balloon-frame cottage on the hillside above the lake. It was all on one floor, around six hundred square feet indoors, with large screened porches on two sides. The inside was unfinished, and it gave the general impression of flimsy minimalism.

We had no plumbing or electricity. Water came from an outdoor iron pump supplied by a well. There was a two-hole privy in back, and a root cellar nearby. Inside were an iron stove, minimal furniture, and one curtained-off bedroom. I can summon up the physical presence of all that with remarkable clarity, 65 years later.

To get there, my parents, brother, sister, and I would pile into our Plymouth and head out of town. The road took us past the First National Bank Building, where we ritualistically said, "Ooh, Ah!" We continued around Indian Mounds Park, where we ritualistically asked if Indians were really buried there. The fact that our parents did not know never dissuaded us from asking the question again the next time. The question, like our admiration of the First National Bank Building, was a part of the trip.

Then we were clear of town and off for a day or two of uncity life. Once we reached the summer place, we shot bows and arrows, looked for hazelnuts in the thicket across the road, explored Wilmes Lake in a rowboat, and spilled water droplets on the red-hot center of that iron stove so we could watch them dance about. When my grandmother joined us, she took me on early-morning walks and told me how to identify birdcalls. Later, I would lie on my cot on the sleeping porch and listen for those same sounds as night embraced me. There were moments in those days when life seemed to be just right.

My older brother met me in St. Paul 40 years later and helped me find the old place. I would never have recognized it. It was surrounded by a housing development. The lake had long since been drained and filled with homes. The most surprising thing about the old cottage was its sturdiness. It was now electrified and had modern plumbing. The wood framing was covered on the inside with plasterboard. This was no longer a summer cottage. It had become a year-round home.

The city had flowed outward and engulfed everything bucolic. It had followed us into the country. Now the country was entirely city, and the city was still moving outward—gnawing its way into other farms and lakes.

As a child of the 1930s, I had seen pictures of titan cities in older books and magazines. But, by the time I was living amidst *Modern*, *Modern* had gone from vertical to horizontal. The automobile had won the hearts of America, and *Modern* no longer meant *monumental*, now it meant *movement*.

Actually, the automobile meant more than just movement; it meant *freedom* of movement. Victorian railroads had, after all, moved people about at modern highway speeds. But rail could never provide the indi-

Wilmes Lake (today's suburban St. Paul, as photographed in the 1930s by my father).

viduality the public demanded. A train constrains one to go just where everyone else in the train is also constrained to go. On the new highways, one went where one chose to go. In the late nineteenth century, one had to use a horse, or a horse-drawn vehicle, to go where one pleased.

Horses and railways remained the only real options until, just at the century's end, four radically new kinds of vehicles began breaking away from this simple dichotomy. They were the bicycle, the motorcycle, the automobile, and the airplane. In one generation's time, these new technologies wrestled with one another for ascendancy while they erased the horse-and-buggy from the American landscape.

I was deeply struck by the speed and completeness with which an epoch in American life had been replaced when I first visited The Carriage Collection in Stony Brook, New York—one of the finest museums of horse-drawn vehicles in the world.[1] The beauty and the variety of the old vehicles astonished me. Both reached far beyond anything I had imagined. I came away wondering what a visitor from the twenty-third century would think upon first visiting a museum of twentieth-century powered vehicles. What a shock that vast variety of Mack trucks, Duesenbergs, Fords, golf carts, and fire engines would be!

So it is with carriages, and, like the powered vehicles in that imagined museum of the future, their variety and diversity had come into being very rapidly. Riding alone on horseback had not given way to widespread travel in wheeled vehicles until Europe began developing extensive road systems. It was the fifteenth century before the closed carriage appeared. It probably came out of the Hungarian town of Kocz (pronounced *coach*), and the name *coach* stuck.

A century later, vehicles had leaf-spring suspensions and better seating arrangements. Still, English carriage builders did not form their first guild until 1677. By the eighteenth century, carriages had begun forcing the development of greatly improved roads. In 1815, MacAdam invented his bituminous macadamized road surface. Now experimentation and adaptation would drive a dazzling array of mutations among horse-drawn wheeled vehicles for carrying people.

The new technologies of coach building were sophisticated. Serious carriage building on this side of the Atlantic didn't begin until shortly before the American Revolution. The first uniquely American rig was the pleasure wagon, a light, basket-like vehicle specifically meant to take picnickers off the beaten tracks. Especially important was the American buggy, which became the nineteenth-century Model T of personal transportation.

Americans built closed coaches as well, but they were more interested in utility than in pomp. I see the greatest beauty in the lightness and buoyancy of America's modest buggies, shays, and phaetons—vehicles for everyman. (A *phaeton* had a light convertible top over a front seat, with an open rumble seat behind.)

Oliver Wendell Holmes celebrated that delicacy in his poem, "The Deacon's Masterpiece, or the Wonderful One-Horse Shay." He begins:

> Have you heard of the wonderful one-hoss shay,
> That was built in such a logical way
> It ran a hundred years to a day,
> And then, of a sudden, it—ah, but stay,
> I'll tell you what happened without delay, . . .[2]

The shay in the poem is designed so perfectly that it lasts a hundred years and then falls into dust all at once. Holmes describes its end:

> What do you think the parson found,
> When he got up and stared around?
> The poor old chaise in a heap or mound,
> As if it had been to the mill and ground!
> You see, of course, if you're not a dunce,
> How it went to pieces all at once,—
> All at once, and nothing first,—
> Just as bubbles do when they burst.

Holmes was a doctor, and his wonderful shay expressed a wish that the human body could work that way, sustaining a lean but graceful quality of life to the very end. He chose as his metaphor the lightness and toughness of nineteenth-century carriage making.

Very few of the notable nineteenth-century buggy-makers made the transition to automobile making, and then survived in the business. The Studebaker Company is the only such name that most of us will recognize. The new internal combustion engines were far more powerful than horses, and they soon devalued the buoyancy and grace of the old carriages. As a child, I saw the last horse-drawn vehicles on my city streets. They were, by then, shabby wagons and trucks still being used to haul ice and produce from house to house. None of the delicate old gigs and shays made it into my childhood. Before I entered that museum in Stony Brook, I had known nothing of all that functional beauty that had vanished during my parents' youth.

Although the automobile seemed to arrive suddenly, we nevertheless keep finding antecedents of the machine, as we trace backward in time. The earliest known steam-powered car was finished as early as 1769 by French inventor Nicolas Cugnot.[3] It was a large, three-wheeled vehicle that moved at the speed of a walk and was meant to haul cannon.

Experimentation with steam automobiles continued all the way through the late eighteenth century and the nineteenth century. As an example, we find a curious article reprinted in America from the British *Spectator* magazine in 1840. The title is "A Ride on Maceroni's Steam-Coach," and in it the author tells of his experience with the machine.[4] The article provides a fine glimpse of the dominant technology of *Modern* in its nineteenth-century cradle.

The author grumbles about having to rise at four in the morning. The ride has to be undertaken before horse traffic appears on the road. The vehicle is large, bulky, and ugly. Its 50-horsepower engine provides more power than it needs. The author wonders why Colonel Maceroni, its owner, did not at least apply some paint to make it look nice.

The gears are designed to move the car at a speed between 12 and 20 miles per hour. It runs up London's Shooter's Hill without any trouble. (Shooter's Hill was still being used as a standard challenge to the new internal-combustion–driven cars in 1900.) The machine does well in that test, but on the way up Maize Hill, it suddenly stops. The author tells about the conversation that follows between the engineer and his assistant:

> "Halloo, what's the matter there?" sung out our engineer: The clutch is broke," was the reply: "Broke! why what is it made of?"—"Cast iron." "*Cast* iron!"—there are no signs by which contempt can be expressed to give effect to the tone this exclamation—"couldn't ye have made it of glass, while ye were about it? Have one of *wrought*-iron made to-day."

For the moment, he is unable to shift gears and he has to *back* the vehicle home. Later, at breakfast, the engineer explains how they were all quite safe riding next to the steam boiler since the water boils in tubes, not in an open chamber. If one tube should rupture, it would not cause a major explosion.

Unidentified nineteenth-century magazine illustration of an embryonic steam automobile.

The conversation between the steam automobile engineer and his passenger continues. They talk about the importance of miniaturizing and silencing steam engines in such a way as to make automobiles a practical and commercial form of transportation.

The automobile would be the eventual outcome of the rising fascination with the new steam engines, but the idea of a horseless carriage had dogged people's minds long before steam. Earlier inventors had tinkered with springs and compressed air. Windmill-powered vehicles had actually been made before that. Leonardo da Vinci sketched self-powered vehicles in the late fifteenth century, Roger Bacon predicted them in the fourteenth century, and even Homer had imagined self-driven vehicles in his writings before the glories of ancient Greece.

Let us then limit our search for the origins of the automobile to vehicles driven by internal-combustion engines and to ones that were actually built. The nominal credit for the first such automobile usually goes to Carl Benz. Benz championed the new internal-combustion engines, and he worked single-mindedly to create a car driven by one. He built a little, three-wheeled motorcar in 1885 and sold his first vehicle two years later. He went into production with a four-wheeled model in 1890, and the Mercedes-Benz Company existed until it only recently became Daimler-Chrysler.[5]

But Benz was not first. The French inventor de Rochas built an auto, and an engine to drive it, in 1862. Two years later, the Austrian Siegfried Markus began working on cars.[6] Markus's second car was rediscovered in 1950. It had been bricked up behind a false wall in the cellar of a Viennese museum to hide it from the Germans. Markus was Jewish, and the Nazis had orders to destroy his car and any literature describing it. When the car was finally rediscovered, it could still be driven.

Markus's story is especially poignant because, if the German Benz believed in the auto, Markus did not. When Markus was invited as guest of honor at the Austrian Auto Club in 1898, he declined. By then Markus called the whole idea of the auto "a senseless waste of time and effort."

Our search, backward in time, for the earliest internal-combustion-driven auto might end in England in 1826. An engineer named Samuel Brown adapted an old Newcomen steam engine to burn gas, and he used it to power his auto up London's Shooter's Hill. Here the whole priority question mires (as it inevitably does) into hair-splitting definitions. What we usually do in these cases is pretty arbitrary. We credit the first commercial success. That's how we credit Edison with the light bulb, Bell with the telephone, and Fulton with steamboat. By that definition, Benz *did* invent the automobile.

The same year that Benz built his first motorcar, 1885, was also the year that the modern safety bicycle (with two wheels of equal size and a chain-and-sprocket drive) came into being. More than the first autos, the safety bicycle fired the public demand for personal vehicles. Generally the date given for the first motorcycle is also 1885. The German builder, Gottlieb Daimler, built a motor-powered bicycle that year, but then he added side wheels to stabilize it. Thus, what he had *really* done was to build an embryonic automobile just after Benz.

By the time Daimler made his pseudo-motorcycle, makers of the earlier and more rudimentary bicycles had long since understood how unnecessary those extra wheels were. Daimler imagined that a two-wheeled vehicle had to be stabilized. Cyclists knew perfectly well that bicycles (like bipedal locomotion itself) are completely unstable and under a kind of constant, easy, second-nature control.

That very instability gives both walking and cycling their extraordinary flexibility. Thus, we can hardly be surprised to learn that the idea of a motor drive for a bicycle had actually arisen immediately after the first modern bicycle, and 17 *years before* Daimler. What *is* surprising is only that the idea failed to catch on immediately.

Writer Allan Girdler tells about Sylvester Roper, born in 1823 in New Hampshire.[7] During the Civil War, Roper worked in the Springfield Armory, where his interest turned to steam power. In 1868, Roper built his first steam-powered motorcycle. The machine was remarkable by any standard. It looked a lot like the new bicycles. It had a small vertical steam boiler under the seat, and the seat itself also served as a water tank. The boiler supplied two little pistons that powered a crank drive on the back wheel. Roper also controlled the steam throttle by twisting the bike's straight handlebar.

Roper went on to build more motorcycles and many steam-powered automobiles. He was far ahead of his time with all his inventions. In fact, he had probably already built his first car during the Civil War. The Stanleys, who built Stanley Steamers, said that they had learned from Roper.

Roper reached the age of 73 in 1896. That June he showed up at a bicycle track near Harvard with a modified motorcycle. They clocked him at a remarkable 40 miles an hour before the machine wobbled and he fell off. Roper was dead when they found him. The autopsy showed that he had died, not from the fall, but from a heart attack, which had very likely been the cause of the wobbling and subsequent fall.

Roper's twist-grip control system, so basic to motorcycles ever after, was reinvented by the early pilot Glen Curtiss— then reinvented yet again at the Indian Motorcycle Company.

An 1886 Roper motorcycle, as shown in *The New Wonder Book of Knowledge,* 1944.[8]

Curtiss, born in 1878, had remarkable mechanical aptitude. He was only seven when the safety bicycle came into being, and, at 22, he (like the Wright Brothers, whom he later battled over airplane patents) opened a bicycle shop. But Curtiss was an ambitious businessman. *He* went on to create his own bicycle *factory*.[9]

He then acquired two engine cylinders and built his own engine. In 1902, he announced plans to start making motorcycles as well as bicycles. Four years later, he was making an avant garde machine with a twin-V engine, along with a twist-grip throttle control.

Despite Roper's early lead, commercial motorcycles were not available until Curtiss and others began producing them at the beginning of the twentieth century. The venerable Indian Motorcycle Company started making motorcycles the same year Curtiss did. Bill Harley and Arthur Davidson built their first motorcycle one year later.

We might well lament how unfair it was that Roper receives no credit for inventing the motorcycle *or* the steam automobile, but the very nature of invention is such that truly inventive people (like Roper) *have to* precede the product-success stories. Any new device that is *not* alien is no more than an extrapolation of things already known. We inevitably reject true invention in its primal atavistic form and wait for people like Benz, Daimler, and Ford to make commercial sense of it. Roper's story can be retold with every major invention that has ever laid hold upon us.

In any case, bicycle makers were the same people who went on to make motorcycles and then airplanes. They sowed the demand for freedom of motion and then veered off into other newer and seemingly freer technologies. Because of the kinship between the people who first developed automobiles and those who built railways, steam cars offered serious competition for internal-combustion-driven autos, well into the new century.

Thus, when we return to the star-crossed question of priority for inventing the automobile, we probably need to follow the swirl of steam backward in time. And that leads us, not to Benz, but all the way back to Cugnot, well over two centuries ago. An old and somewhat conservative thread weaves through the whole story of the automobile. The bicycle represented a populist movement that seized the public's imagination immediately. But it was after World War I that the automobile fully laid its hold upon the general public.

Historian Kevin Borg casts a fine light on tensions between egalitarianism and elitism surrounding the automobile when he talks about a 1906 headline in the *New York Times*: "Chauffeurs Lord It Over Their Employers." That headline typified a rising problem in the first days of the automobile. The wealthy were having a great deal of trouble with a new breed of servant.[10]

Before the twentieth century, only the wealthy owned the fancy sort of carriage that was driven by a coachman. A separate staff of servants would typically care for the horses and equipment and do the driving as well. When automobiles first came on the market, it was those same wealthy owners who could afford them. The coachmen thus became the natural heirs to the care and handling of autos, and they could not have been more wrong for the job.

The first automobiles needed constant repair. In 1903, Cadillac advertised, "When you buy a Cadillac, you buy a round trip." That meant that a Cadillac was apt to get you home before it broke down. Even then, the new coachmen absolutely had to have the skills of a mechanic.

The role of the head coachman was, by then, well defined. Toward his superiors, he was deferential and obedient, although he skimmed the salaries of the servants whom he managed as well as the money he received for supplies. He had neither the talents nor the temperament of a mechanic. Those traits obviously had to be replaced by a new kind of employee.

The people who entered the twentieth century with a mechanic's knowledge generally had the egalitarian instincts that had created automobiles in the first place. They had no sense of the old social stratification, and they were not about to assume the role of a nineteenth-century servant. They ran roughshod over their new bosses and went joyriding in the expensive cars in their care. Of course, they too skimmed their operations money, but they did so without the veneer of social decorum. They were in control, for who else could deal with those terribly complex new machines?

By the end of World War I, two things had happened with automobiles. They had stopped needing constant repair, and their owners had learned to drive them. Chauffeurs did not vanish; they merely lost their power. The new boss/employee relationship began to stabilize.

That same drama plays out in any new technology. I spent years at the mercy of technical typists. They were underpaid, but they were in control. When word processors came on the market, I was first in line to buy one. It gave me my ticket out of a relationship that degraded everyone.

A school for the new "class" of chauffeurs, from *The American Review of Reviews*, 1909.[11]

During the past 20 years, computers and the people who know how to use them have replaced the typist and the file clerk. Power has devolved into the hands of whoever really understands those computers.

Sometimes those are the people in the data-entry trenches; sometimes it is their bosses. In large organizations one finds computer specialists who are in danger of falling into the same role as those early chauffeurs. Knowledge is power, all right, but it is always temporary power. In the early twentieth century, we needed to relearn the old lesson that knowledge flows; what one person knows, another can also know. During the years before World War I, we needed to reduce knowledge of the automobile to the commonplace, just as we had to do a century later with knowledge of the computer.

One powerful medium by which motor vehicles were being brought before a larger public, and gaining their confidence, was that of races and competitions. In the summer of 1905, a 1904 Oldsmobile took 44 days to make it from Hellgate in New York City to Portland, Oregon. Four years later, a race was held in which an early Ford made the trip from New York to Seattle in only 23 days. And a new game was afoot.

The most amazing of these early races occurred two years before the first trans-America race and was far more ambitious. In 1907, a Paris newspaper announced an *around-the-world* automobile race. This, at a time when automobiles were still essentially handmade machines running on roads intended for horses and horse-drawn vehicles!

Writer J. M. Fenster tells how the paper specified a route for the race.[12] The race was to proceed westward from New York to San Francisco. Then a steamer link would carry the cars to Valdez, Alaska. They would continue driving across Alaska. Another steamer would ferry them over the Bering Strait. From there, the cars would make the daunting trip across Asia and Europe to Paris.

Six cars accepted the challenge. One each came from Germany, Italy, and the United States. France fielded three cars. The 20,000-mile race began in Times Square on February 11, 1908. The organizers had hoped that this starting date would allow the cars to make it across Siberia before it melted into summer mud. Each car carried three or four riders, piles of gear, and two hundred or so gallons of gasoline. Spare parts and fuel were to be brought along by railroad train or dogsled, depending on where the cars were.

Crossing New York in the winter proved to be the first severe test. A French car gave up in a snow bank after only 44 miles. The American car, a Thomas Flyer, shoveled ahead. The German team, led by a serious army officer under the scrutiny of the Kaiser himself, soon lagged far behind. People at each whistle stop celebrated as contestants arrived. Then they charged outrageous prices for food and lodging.

The Americans got to Alaska first and found that conditions were impossible. After an exchange of telegrams, officials changed the route. Now the cars were to get from Seattle to Yokohama by ship. They were then to cross Japan and catch another ship to Vladivostok.

The cars finally gathered at Vladivostok, waiting to be restarted in the order in which they had arrived. The Germans had made part of the American crossing on a train. As a penalty, they were to start 15 days after the Americans. But they bolted ahead. Other teams hesitated when they heard stories of bandits in China. One dropped out.

The Americans finally reached Vladivostok and took off after the Germans. They drove like men possessed, while the Germans drove along railway beds in violation of the rules. The Germans finally reached Paris in July, after 165 days on the road. They were hailed winners until the Americans arrived four days later. Then the Americans were hailed as the winners. The naming of a winner did not really matter, since there was no purse. The real accomplishment was simply reaching Paris. The race finally ended when the Italians did so that fall.

I first drove across America 46 years later, and it took me only four days. The great American Interstate Highway System had yet to be authorized, much less built. However I did most of the trip on the broad-shouldered, two-lane highways that had come into being only because of theater, like this now-forgotten, and seemingly impossible, automobile race. One surely has to wonder what to make of a race that averaged little over a hundred miles a day. But before people would invest in roads, they needed to know that those rugged new machines could go anywhere. And before the public would seriously invest in automobiles themselves, they had to know that owners could use them *without* hiring a chauffeur. The race was, in that sense, a major step on the way to inventing Henry Ford.

Ford was already on the scene. He began producing the Model T about the time the race was completed in 1908. By 1914 he was pouring out cheap, conservative transportation for America using his new assembly-line methods. Ford was quoted (probably wrongly) as having said, "You can paint it any color, as long as it's black." The Model T appeared 29 years after an inventor named George B. Selden had filed the first important American patent for a combustion-powered automobile in 1879.

Selden was a Civil War veteran. After the war he studied engineering at Yale, where the great American scientist J. Willard Gibbs was one of his teachers. Selden had to drop out when his father died, so he studied law and passed the bar exam in 1871. Selden supported himself as a lawyer while he worked long hours on the creation of his motorcar. He had to create his own version of the clumsy two-stroke Brayton engine, but he nevertheless had an automobile running by 1878.[13, 14]

George B. Selden in his "Benzine Buggy,"
from *The Gasoline Automobile*, 1920.

Selden was aware of several things that might have gotten past another inventor: He knew that his patent could protect him for only 17
years once it had been issued. He knew that it was unlikely that he could
produce cars and create a market for them that soon. He certainly understood that his own abilities as both an inventor and a lawyer far outstripped his talent as a production engineer.

There is also evidence that he increasingly felt the injustice of having
produced the seminal automobile and then having suffered the public
ridicule that accompanies any such seminal achievement. That may well
have been a major factor in what followed. Selden kept his patent alive
by filing amendments that delayed its issue, while the Duryea brothers,
Olds (of the Oldsmobile), and many others created workable cars. Duryea
cars were on the market while Selden was still struggling to produce a
workable model.

Selden's patent nevertheless surfaced between 1901 and 1903 as a road-
block to the successful makers. One after another, they came to terms
with it and paid Selden a 1.25 percent royalty on the retail price of each
car they sold. The Selden organization was called the Association of Licensed Automobile Manufacturers, or ALAM.

In 1903 the ALAM denied Ford a license to build cars on the basis
that he was only an assembler, not a manufacturer. Just how great an
assembler he was did not become clear for another decade. However, in
a meeting in 1903, Ford's business manager, James Couzens, told the

ALAM, "Tell Selden to take his patent and go to Hell with it." Ford, sitting back in his chair, added laconically, "Couzens has answered you."

So Ford and his people took to the courts where Ford painted the Selden patent-holders as a great corporate trust that was trying to crush him. He finally won his case in 1911. By then, the 30-year-old Selden patent looked pretty antediluvian. Ford had already made 20 times as much money as the so-called Selden monopoly and that was *still* before he created the modern assembly line.

Olds had been the first to mass-produce cars using interchangeable parts. Five years after his patent victory over Selden, Ford adapted and expanded upon Olds's methods. Ford then started to make money on a scale no one had ever imagined. Selden was forgotten, Ford turned into an American legend, and the Model T finally completed the creation of free movement that the safety bicycle had begun almost 30 years before.

Thus motor vehicles at last turned from a novelty for the wealthy into a primary instrument of American transportation. In America, fewer than four hundred thousand had been built by 1909, the year after Ford introduced the Model T. In 1920 that number was well over 11 million. Yet cars remained very hard to operate. One had to keep both hands and both feet in action all the time. Gear shifting and steering both demanded significant upper body strength. Furthermore, a full generation of drivers had to use a hand crank to start their cars. That could be murderously difficult on a cold day.

Consequently, it was a rather remarkable woman who undertook to drive any of the early motor vehicles. We gain some insight into those difficulties if we read, of all things, Gertrude Stein's autobiography, told in the voice of her lifelong companion, Alice B. Toklas.[15] Toklas and Stein were living in France when World War I began. Late in the war, they took on the assignment of driving a Red Cross ambulance. And an odd pair they made:

Stein couldn't deal with military officialdom, and Toklas couldn't deal with the vehicle. Stein had both the mechanical talent and the muscle. She could field-strip the primitive ambulance, and (although her driving was notoriously hair-raising) she loved doing it. So Stein and Toklas fixed upon a division of labor. Toklas dealt with the military while Stein did all the driving and maintenance.

Stein struck up a special friendship with two American mechanics who teased her every time the ambulance made any noise. "That French chauffeur is just shifting gears," they would say. That remark went to the heart of a primary problem in those old machines. Not only did they lack automatic transmissions and power steering, they also lacked the synchromesh gears I took for granted during all my years driving stick-shift automobiles. To shift these earlier vehicles, one had to double-clutch and catch the moving gears at just the right point to avoid stripping them. That often required an athlete's strength and skill.

The Hurry Call: The Night of May 30, 1918,
from a drawing by Captain Harry E. Townsend,
The Century Magazine, March 1919, p. 655.

Early during World War I, an American engineer, William C. Stevens, heard about women ambulance drivers. He designed electric motor control systems, and he knew how clumsy all gearshifts were. Stevens realized how driving could test a woman's upper body strength. If Stevens could simplify shifting, he would almost double the number of people who could drive motor vehicles.

So he used what he had learned from electric motor systems and developed his complex "preselective gear-shift." A box on the steering column had five buttons: Neutral; Reverse; and Gears 1, 2, and 3. The driver would push the button, shove in the clutch, and the gears would shift.

As early as 1916, two now-forgotten cars used Stevens's invention. They were the Apperson and the Premier. By the early 1930s, Oldsmobile had made the first Hydromatic system. It was the precursor of the automatic transmission. A few years later, the Bendix company made the precursor of the synchromesh system, which became basic to manual transmissions. Stevens's push-button systems made a brief comeback in the 1950s.

When I drove army trucks in 1955, gear-shifting problems were far from solved. I had terrible trouble with the standard ton-and-a-half truck. The experience leaves me filled with admiration for the adroit and practical Gertrude Stein and with great respect for the complexity of a problem that dogged automobile driving during much of the twentieth century.

One can read this laboring evolution in old automobile advertisements. They talk about the difficulties and explain how this car or that is easier to use than others. The ads focus on the machine, not the user, and women simply do not appear in them. Historian Pamela Walker Laird traces early ads and finds a remarkable absence of personal enthusiasm or hyperbole.[16]

The first car ads came out just after P. T. Barnum had emerged on the American scene and the new safety bicycle had swept the country. Bicycle sales were fueled by ads that said things like, "Columbia riders know naught but pleasure." Never mind the specific virtues of the Columbia bike; its advertisers knew that people would be drawn in by the pleasure bicycling offered.

Early automobile drivers wrote about how "The best part of automobiling . . . is the way [you sweep] uphill [and] make gravitation slink away crestfallen, conquered." But that kind of language did not make it into advertising copy.

Ads for the 1922 Dorris explained its engine and its choke. They showed a diagram of its fuel system. All that did help teach the public how this new part of American culture worked. But to promote their car, the Dorris people said something one would never find in an advertisement today. "The Car Without a Single Weakness," one banner proclaimed.

Weaknesses, Laird points out, are the last thing any advertising person wants the public to be thinking about. They are, instead, what designers think about all the time. "How can I keep the rings from wearing out, the suspension stable, the radiator cool?" Early auto ads, tightly controlled by people who made autos, were terribly self-referenced. They spoke to the concerns of the company, not to those of the buyer.

Early ads also played the angle of prestige. They announced that their buyers were upscale. No Barnum tactics for them! Car ads showed a full view of a static car. People, if any were in the picture at all, were wealthy admirers. "Autocrats of the Road" cried an early Oldsmobile ad. "Ask the

To start the Triumphant New Studebakers you simply switch on the ignition with a key. The engine instantly responds—and even should it stall at any time, it automatically starts again.

The Studebaker Synchronized Shift assures instantaneous, silent shifting in all gears and at any car speed. There's no clashing. You shift as fast or as slowly as you wish.

The improved Studebaker brakes are adequate to any emergency of road or traffic. Brake drums are larger. Lining that's molded and thicker doubles the life and halves the wear.

Women love to drive these
triumphant new Studebakers

There's a grace to the Air-Curve Coachcraft of these Triumphant New Studebakers that only half suggests the commodious interior comfort of larger, wider bodies. Driving seats are instantly adjustable and very capacious. No American cars have roomier rear seats than the President and Commander Sedans—55 inches across. The Dictator Eight, and the new 117-inch Studebaker Six, have proportionately generous seat dimensions. Drastically lower in price, these Triumphant New Studebakers present a total of 32 startling betterments, chief of which is vastly finer Free Wheeling plus new *fast-action* Synchronized Shifting.

In this Studebaker advertisement from the February 1932 *House & Garden* magazine, the company strives to reach female drivers. It boasts an important early simplification of gear-shifting, which it calls "Free Wheeling plus new fast-action Synchronized Shifting."

man who owns one," said Packard, with the sly hint that that man might be someone above your social stratum.

Historian Virginia Scharff captures what was going on here wonderfully well in one chapter of her study of women and early automobiles, *Taking the Wheel.* Its title is "Corporate Masculinity and the 'Feminine' Market."[17] She describes the industry's learning pains. Henry Ford, in particular, doggedly kept selling to everyman; he kept the tone of corporate self-reference long after others got the message. He showed his own factory smokestacks, and the copy said, "A Giant Who Works for You." Buyers don't care how the product came into being, Laird says, they are interested in the result.

During the 1920s, car ads finally reached the hands of advertising professionals, separate from manufacturers. Once they did, we find the Lexington automobile company announcing that it was "Built to Stay Young." In a pivotal new ad, we see a girl in her Jordan automobile racing a cowboy on horseback. The cars now *move*, and it is no longer just well-to-do men who drive them.

Still, not everyone got the message. In 1931 we see a parked black Ford with the message, "The New Ford is an economical car to own and drive."

By then, General Motors, and Chrysler Corporation, and others were becoming fully competitive in their attempts to claim a broad popular

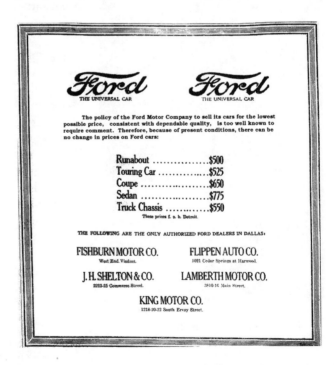

Take-it-or-leave-it Ford advertisement from the *Dallas Morning News,* Jan. 18, 1919.

A SEDAN OF COMMANDING BEAUTY
for Men in Command of Affairs

One can see the shift in mood of advertising in two images from the 1926 *Saturday Evening Post*. Studebaker (left) is still appealing to upper-class males. Jordan (right) is openly hedonistic and trying to reach men and women alike.

market. And other companies would now display their cars speeding, carefree, down some country road.

Ford, however, did have his own unique sense of the marketplace and how to play it. What he did in 1927 offers a splendid example of a kind of merchandising not seen again until Bill Gates repeated his scenario. It will help us understand just what Ford did if we first think about the prerelease advertising for Windows 95.

Microsoft engineered a spectacular merchandizing blitz before the release of Windows 95. Consider the choreography. They gave the new product a name and a logo. Everyone in America who had used earlier versions of Windows knew that a major change was coming. Microsoft advertised Windows 95 heavily and delayed delivery. Bugs had to be found before it could be sold. Selected users across America test-drove Windows 95 on their computers in its beta version, and they reported trouble back to Microsoft. Tension built as the program was fine-tuned. Finally, delivery day! Sales flew off the charts.

Now let us move back 68 years. Automotive historian Michael Lamm observes that what we saw with Windows 95 was only a pale ghost of the panache that Ford showed us in 1927.[18] Henry Ford had been making Model Ts since 1908 with no major changes, and this car had changed the face of America. It had been the standard of cheap, effective transportation. Other automakers were making improvements that were leaving it behind, but the Model T was still a major competitor.

Ford had to pull a new rabbit out his hat if he meant to stay on top, so he announced a new car. He did not name it the Model U to follow T. Rather he went all the way back to the front of the alphabet and called it *Model A*. This was to be a wholly new beginning in automobile making.

Lamm tells what happened next: In May 1927, the fifteen-millionth Model T came off the production line, and Ford abruptly ended its production to retool for the Model A. That proved to be a brilliant (if, perhaps, unintended) public-relations ploy. Suddenly Ford's assembly lines simply stopped, and dealers all over America had no cars to sell.

America waited. Ford said that information about the new Model A would follow in a few weeks. June passed with no word from the Ford Company. The media printed would-be pictures of the new car. Some said it would have a six-cylinder engine; some said eight. The venerable Model T planetary-gear transmission was certain to be replaced by something with a modern gearshift. Beyond that the public knew nothing.

Dealers had no cars to sell, yet only a few defected to Chevrolet. America waited. Finally, in November, Ford announced that the Model A would hit the market on the second of December. By then, the first 125,000 Model As had been sold sight unseen.

When the car finally went on display in Madison Square Garden, over a million people came to see it on the first day. Three-quarters of a million orders were placed within the first six weeks, and Ford stayed backlogged until well into the following year. Model A replaced Model T as the new automotive legend, and, regardless of a fluid marketplace, Henry Ford and his car continued to touch a deep-rooted populist chord. They mutated into the stuff of folklore.

Five years later, in 1932, the Ford V-8 arrived. It was a fine automobile and the public truly loved it. Soon after it came on the market, the popular song, "I'm an Old Cowhand from the Rio Grande" appeared. That one

Early Model A convertible in a Model A parade at Mystic, Connecticut, on Sept. 21, 2001. (In the background is the famous *Mystic Pizza* parlor.) (Photo by John Lienhard)

had us all singing: "I'm a riding fool who is up to date . . . Cause I ride the range in a Ford V-8."

Ford received a letter in 1934, purportedly from a notorious user of the Ford V-8. He was Clyde Barrow, of Bonnie and Clyde fame, and he had taken time out from robbing banks to identify his name with Ford's. The text of the letter was as follows:

Dear Sir:

While I still have got breath in my lungs I will tell you what a dandy car you make. I have drove Fords exclusively when I could get away with one. For sustained speed and freedom from trouble the Ford has got ever other car skinned and even if my business hasent been strickly legal it don't hurt enything to tell you what a fine car you got in the V-8.

Yours truly
Clyde Champion Barrow.

The Barrow family opined that the letter was a forgery. The handwriting didn't look right, and Clyde would have been unlikely to use his full name in the signature.[19] However, the letter accurately represented Clyde's feelings toward the Ford V-8. He really did steal V-8s almost exclusively, and police later found that the V-8's steel construction had given him better protection from bullets than other cars.[20] The V-8 may well have saved Bonnie and Clyde as they sped away from police firing after them with conventional side arms.

Of course, the side of a stationary Ford V-8 sedan could never have saved Bonnie and Clyde from that fusillade of 30-06 fire from point-blank range on May 23, 1934, when they were finally caught in an ambush. However, the very existence of such a letter (real or forged) expressed the grassroots appeal that Henry Ford had managed to harness.

Two issues swirled about the role of the automobile of the 1930s. One was egalitarianism; the other was speed. The issue of speed eventually closed in upon the automobile, while the issue of egalitarianism seems to open into infinity. The problem with speed was that it became simple, self-limiting excess.

We had a phrase we often used back in the 1930s: "Boy, she was goin' like 60!" That meant 60 miles an hour. The speed of one mile a minute had a certain cachet. We called the nonstop train that ran from St. Paul to Chicago "The 400" because it supposedly made the four-hundred-mile trip in four hundred minutes. That was speed! We had, of course, seen much higher speeds by then. A train passed one hundred miles per hour in 1893, an automobile passed it in 1906, and airplanes finally went that fast during World War I.

Automobiles promised speed for everyone as soon as roads caught up with them (more on that matter in chapter 9). I will always remember the

Race car illustration, from a painting by William H. Foster,
June 1913 *Century Magazine*, p. 219.

thrill of fear when my father pushed our old Plymouth over 60. We were
riding on the best road around when we did it—a two-lane tarred highway.

Henry Ford had used a frozen lake to reach 91 MPH in 1904. The Stanley
brothers drove their steamer 128 MPH on the packed sand of Daytona
Beach in 1906. Soon after, most states enacted laws against road racing.
Speed had become the great American hunger. Writer Michael Gianturco
tells about a lost technology that spoke to that new hunger. It was the
wooden racetrack.[21]

In 1910, a Barnum-like man named Jack Prince sold an investor on
the idea of building a wooden speedway at Playa del Rey in California.
The famous racecar driver Barney Oldfield immediately did a lap at the
amazing speed of 99 miles per hour. The Playa del Rey track was made
from thousands of boards held together by tons of nails. Prince's first
design was a one-mile circular bowl with steep, sloped sides.

A fascinated public watched as fatalities mounted. Finally, in 1913, a
fire destroyed the track, and Damon Runyan wrote, "Playa del Rey burned
last night with a great saving of lives." But the mania had only begun.
Wooden tracks grew like weeds. By the 1920s, curves had been banked
until they were unsafe at anything under a hundred MPH.

The brick Indianapolis Speedway was built at the same time as the Playa
del Ray track, and the Indy is considerably slower. The Indy 500 has always
been about distance as much as speed. In 1926, a car lapped a wooden
track in Miami at 143 MPH. The fastest lap that same car could do at Indy
was only 112 MPH. But wooden tracks were an excess of the pre-Depres-
sion years that ended abruptly when the economy failed. Some were burned
for firewood, while cars became a commonplace part of American life.

The quest for speed has continued right up to and beyond the speed
of sound, but now that quest occurs only in solo, straight runs on very
flat ground. The 2001 winner of the Indy 500 averaged only 153.6 miles
an hour, less than *one-quarter* of the speed of sound. Indy cars still reach
only 30 percent of Mach-1 on straight runs.

Real land transport, even track racing and high-speed rail service, seldom gets beyond 150 or 200 miles an hour. Behind all this is the simple law of physics, which tells us that kinetic energy increases as the square of velocity. If you have ever suffered a car accident at even 25 miles per hour, you've been through a horrific experience. At top racecar speeds, the violence of a collision is a hundred times worse, at the speed of sound it's a thousand times as violent, and at orbital velocities it is a million times more destructive than a typical fender-bender accident.

Take those road turnouts on highways coming out of the mountains. Those spurs go sharply uphill to catch a truck if its brakes fail. Consider how far uphill you would coast if those spurs were not filled with sand to slow you down: At 25 miles an hour you'd stop after you'd risen 21 feet. At 200 miles an hour, the road would have to rise a quarter of a mile. But, at the speed of sound, you could coast from sea level to the top of Mt. McKinley!

The increase of damage with speed is only one limitation to speed. There is also the driver's ability to respond. At the speed of sound (five seconds per mile), the best driver will go a long city block in the time it takes to perceive danger and react. No one even thinks about racing at such a speed. Indy 500 drivers take the better part of a football field to react. Yet they constantly come within a few feet of each other, and accidents are frequent. On the highway, you and I need at least 70 feet just to react. Small wonder that highway deaths rise when we lift speed limits.

The reaction-time problem can be removed by giving control over to computers. That is already being done in airplanes and high-speed rail systems, though not yet in cars. But the sharp increase of damage with speed remains. A pothole or a bent rail, a blowout or a rock rolling onto the road, will keep making high speeds unattainable on land. Our grand-children are not apt to see much ground travel above two hundred miles an hour in their lifetimes.

Of course, the terrible lure of speed reaches the young in particular. I have never again driven as fast as I drove back in 1946. But then neither have Americans in general driven significantly faster than we did back then. Speed was an elixir of *Modern* that simply could not be increased any further after the 1930s.

Luxury was another elixir that constantly affected our view of the new automobiles. Advertisers tried to promote luxury when they tried to sell modestly priced automobiles to John Q. Public. But once the con-straint of economy was released, the results were dazzling.

Clark Gable's favorite possession was a 1935 Duesenberg SJ coupe. It weighed two tons with a 250-horsepower straight-8 engine. It could go 140 MPH, and it was the finest thing on wheels. A buyer had to first pur-chase the Duesenberg motor and chassis for $8,500 and then go to a coachbuilder for a custom-made body. It came to a huge sum back then, and auto historian Brock Yates tells us that collectors will now pay $3 million for one of those old Duesenbergs.[22]

In 1900, Iowa bicycle makers August and Fred Duesenberg had begun playing with gasoline engines. When other bicycle makers did that, they went toward flight. But in 1906, August and Fred got money from Edward Mason, an Iowa lawyer, to manufacture cars. Frank Maytag (of washing-machine fame) bought Mason out. For a while, the Duesenbergs built the Maytag-Mason automobile.

Neither Maytag nor Mason had the hang of the car business. The company gradually folded while the Duesenbergs went to St. Paul to work on engines. By World War I their engines had made a good showing in the Indy 500. Eddie Rickenbacker drove cars powered by those motors before he flew in the war.

The Duesenbergs made airplane engines for the Army Air Service during World War I, and afterward they made a car under their own name. It had the best engine, but they couldn't get the body right. Worse yet, they began competing with a California car maker named Miller.

Both companies dreamt of making cars that would both serve the public and compete on racetracks. That may have been poor thinking, but Yates points out that it led to furious cross-fertilization. Auto design profited even if the auto business did not. Speeds rose, and America has yet to build highways for machines like the Duesenberg.

Then another maker of fine autos, Errett Cord, joined the Duesenbergs in 1926 to make the luxury J-model Duesenbergs. From 1928 to 1937, they made 481 of those glorious cars and sold them to the rich and famous—to the Maharaja of Idore and Gary Cooper. No carriage trade is likely to sustain itself indefinitely at such a level. The company went bankrupt in 1937. The cars had been an unwise business practice, but their craftsmanship was superb. In the end, such love of invention could not fail entirely. Duesenbergs had set the pace, driven the industry, and touched America.

They also shaped our metaphors. Cartoons still use that long in-line 8-cylinder engine as the very symbol of automobile excess. The Duesenbergs inadvertently added a slang term to the English language. The word was *doozy*, as in, "Isn't that car over there a real doozy!" What each of us meant to have was a 200 MPH Duesenberg that we could buy for four months' pay at our job in the local factory.

So the automobile delivered its conflicting messages of personal freedom, elitism, populism, luxury, economy, speed, and general excess. But, as the American public sorted those messages, it worked upon the culture in ways that no one could ever have anticipated. Let us look next at the surrealistic ways in which that worked.

9

On the Road:
Of Highways and Gasoline

Any massive infusion of new technology brings with it all kinds of unintended consequences. Certainly, no automobile maker was thinking about ending the impulse toward building titan cities. And the outward diffusion of urban life was only one of many effects brought on by the automobile. Beneath the obvious rearrangement of our dwellings ran a parallel reshaping of our metaphors, our language, and the very ways in which we thought about the world around us.

We saw in chapter 8 that advertising provides a fine window into the dimensions of *affect* that weave themselves around any technology. Advertisers have always used a trial-and-error process that eventually finds means for breaking through our rational barriers to awaken fear, hope, and desire. Now and then they do so very effectively; and when they do, we always gain new insights into ourselves.

Advertising is much like pun-making. It is a kind of human ingenuity that everyone loves to hate. When advertising is good, it becomes true folklore. On the one hand, God help us if we cannot ignore advertising. On the other, we would be much the poorer if we did not allow ourselves to be informed by it.

Starting in the 1920s, advertisers began to speak directly to passing automobiles. By the mid- to late 1930s, my family was making great automobile vacation trips through the American West each year. Many of my strongest early memories are recollections of the constant conversation that went on between the car and the roadside.

We would grind our way along concrete, asphalt, or gravel roads through the barren Dakotas, rugged Wyoming, and bourgeoning California. We would count windmills on the endless plains of Nebraska, or cling to our seat on hairpin roads through the mountains of Colorado. We would find lodging each night for $1.50 or so in a one-room cottage with a stove and a detached privy.

As we wandered across the sprawling West, those roads held one very particular delight. Every hour or so we would pass a string of six red signs, each with a few words of doggerel.

SHIVER MY TIMBERS,
SAID CAPTAIN MACK,
WE'RE TEN KNOTS OUT,
BUT WE'RE TURNING BACK,
I FORGOT MY,
BURMA-SHAVE.

Burma-Shave signs were as surely the mark of the American landscape as windmills, barns, and purple mountains in the distance. They delighted us from 1927 until 1963, when the Odell family, who owned Burma-Shave, sold the company. The product was taken over by Philip Morris, and the signs, with their good-humored verve, soon disappeared.

Grandpa Odell, patriarch of the Minneapolis-based Odell family, was a lawyer who sold liniment on the side.[1] He told people that a sea captain had given him the recipe. That could well have been true, for sea captains once did traffic in nostrums from exotic lands.

Then Odell's son Clinton formed a company and became serious about selling liniment. Clinton eventually suggested to his two sons, Leonard and Allan, that they might create a brushless shaving cream. Leonard and Allan worked with a company chemist, and they stirred up some three hundred recipes. They finally made one that really worked. They called it Burma-Shave, and they took it to market in 1925.

The product did poorly at first. Then, in 1927, Allan thought of using roadside signs. The first set of signs achieved nothing close to the rhythm of the later ones. This set said:

SHAVE THE MODERN WAY
NO BRUSH
NO LATHER
NO RUB-IN
BIG TUBE 35 CENTS DRUG STORES
BURMA-SHAVE

All that changed in 1929. Burma-Shave developed its wonderful six-sign doggerel, kinky humor, and surprise endings. Here is a particularly nice sequence of puns:

IF HARMONY
IS WHAT
YOU CRAVE
THEN GET
A TUBA
BURMA-SHAVE

Those playful red signs helped teach kids like me to read. They also may well have saved lives with their catchy safety messages.

> HER CHARIOT
> RACED 80 PER
> THEY HAULED AWAY
> WHAT HAD
> BEN HUR
> *BURMA-SHAVE*

At the very least, with the signs spaced a hundred feet apart, speeding cars regularly slowed down to read them.

The advertising certainly succeeded. Today, if you're not an electric-shaver user, you have very likely adopted brushless shaving creams. But the other lasting effect of the signs is the peculiar ingrained meaning that they have for anyone born before the mid-fifties. Readers "of a certain age" will surely smile at this one:

> IF YOU DON'T KNOW
> WHOSE SIGNS
> THESE ARE
> YOU CAN'T HAVE
> DRIVEN VERY FAR
> *BURMA SHAVE*

The almost universal nostalgia raised by those funny red signs reflects the way in which we were being linked with one another as we had never been. An odd synergy surrounded the automobile and the new culture of the open road. If we did not happen to have a car at any given moment, then we hitch-hiked. It was a favor we would later repay.

A shorthand of signage thus grew up along the highways. Language itself became foreshortened. Burma-Shave signs were one very evident part of it, but all kinds of messages reached out to catch the new automobiles as they passed.

Lydia Pinkham's Vegetable Compound had been around since 1875. During the 1920s, drivers began seeing it widely promoted by huge signs painted on the sides of barns. It was an herbal compound that was, until 1960, stabilized with 18 percent alcohol. The medicine was claimed as a cure for early cancer, tumors, flatulence, and much more. It quite possibly *was* somewhat effective in easing menstrual cramps. Although the alcohol was identified as a "solvent and preservative" only, Lydia Pinkham's palliative predictably grew in popularity as an acceptable way around Prohibition in the 1920s and '30s.

Signs beckoned us in to roadside menageries, to fruit stands, and to the new motor inns. For a thousand miles on either side of Wall's South

Dakota drugstore, signs on the road told about that kitsch oasis far ahead. (Those particular signs still stand today; you may well have seen them.)

Signs would eventually become a blight that had to be controlled, but that would be in some unforeseeable future. For now, signs spoke to our cars, and not just signs. Actually, an entire architectural movement spoke to them as well. It was a movement that arrived shortly before I was born and somehow disappeared much later in my life, when I wasn't looking. A strange new architecture of instant communication first emerged out of California.

One began finding that one could buy lemonade from a building shaped like a lemon, ice cream from an igloo, or film from a shop shaped like a camera. Buildings also took forms without rhyme or reason—diners shaped like boxing gloves, Zeppelins, dogs, pumpkins, and tee-pees (as in the now-long-abandoned Teepee Motel down in Wharton County, Texas). These were pure means for catching a driver's eye, all glorious in their frivolity.

Architecture is supposed to be subtler than that. It should tell its function indirectly. However, one has little time for subtlety at 60 miles an hour. So advertisers did what medieval businesses once did for a different reason. A medieval innkeeper did not write the name of his tavern, "Head of the Horse," over the door. Too few people could read. Instead he displayed the carved head of the horse itself.

California rediscovered that childlike directness. The state had already declared its architectural independence with Spanish colonial adobe and stucco. Next it moved on to exotic themes—Grauman's Chinese Theatre and the Aztec Hotel. California was reaching for the splendor of the silent movies.

Ruins of the Teepee Motel, on a road left behind, with the rusting sign shown at right. (Photos by John Lienhard)

This decaying relic of a B-17 announces the Bomber diner in Milwaukee, Oregon.
It is a lovely anachronism. Put up just after World War II, it was a last gasp of
mimetic architecture. Yet it still stands today. (Photo by John Lienhard, July 2002)

By the 1930s, California had moved all the way to Wonderland. A
1936 Los Angeles Coca-Cola Company building was shaped like a great
ocean liner. You entered a huge coffee pot to drink coffee. Sometimes
form followed function as doggedly as your shadow follows you. Sometimes function was as irrelevant as a daydream. It is called *mimetic*, or
vernacular, architecture, and it was far more than mere cuteness.[2] It was
a way to make the break so we could begin again. The new cars told us
that we had to reorder our lives, and we answered in a new language
along the side of the road.

Into all this was intimately woven another interaction between road
and automobile, one that was initially wed to mimetic architecture but
that has outlived it. It is a connection that came home to me with particular force one evening, not too long ago, when my wife picked me up
after work at the university.

"Where shall we eat?" she asked. We ran through our usual set of fast-
food shops and finally set out, with the unspoken agreement of two lem-
mings, for an upscale hamburger emporium.

You see, I had just found a book on the history of the White Tower
chain, and we had paused to turn its pages before we drove away.[3] If you
are old enough, you too will remember those 5¢ hamburgers once served
up at the old White Castle and White Tower establishments. Sunday evenings, my father would pick up a bag of White Castle hamburgers—
small, fried in sizzling water and onions along with their own fat, and
garnished with pickle slices. We would eat them while we listened to
Jack Benny on the radio.

The White Castle chain had been founded in Wichita in 1921 and the White Tower chain in Milwaukee five years later. These restaurants were exercises in minimalism. Both were housed in small, radiant white buildings with a medieval tower for an entrance. Inside were gleaming chrome and white tile. The menu was as narrow as a slat; White Tower offered hamburgers, coffee, ham sandwiches, pie, donuts, and soda pop—mustard but no ketchup. The food was all served on paper napkins.

Those tiny buildings were the purest mimetic architectural kitsch of the 1920s—a slightly exotic medieval theme, with simple, eye-catching glitz to draw in the new automobile trade. But, if they were creatures of the '20s boom, they had two qualities that gave them staying power for the Depression years of my childhood.

One was Spartan simplicity. They were the minimal means by which to feed people who had only a nickel. The other feature was that dreamy veneer of medieval fantasy. A place to escape the hard reality of hunger that was all around. If you saw the Susan Sarandon movie *White Palace*, that dreamlike simplicity was now gone—lost in a postmodern version of the old White Castle counters.

Money and hope had evaporated in the 1930s, while the two chains built and expanded. Cheap food and cheap entertainment are commodities that survive in the worst of times. Fifteen-cent movies, five-cent hamburgers, and Jack Benny on the radio. That was the stuff of my childhood, back in some of America's hardest days.

And so, my wife and I, drawn by those primal memories, ate our large, thick, postmodern hamburgers with side servings of garlic couscous. We paid 120 times what we had once paid the White Castle. We ate and talked about that sweet, impoverished world between two wars—about wonderful five-cent pleasures in kitschy little fairy-tale diners.

It seems hard to believe that what we recalled with such fondness was no more than the technology of barely getting by in the worst of times. While mimetic architecture has long since faded, the automobile-driven consumption of fast food has not only grown; in many of its forms it has also begun posing a postmodern assault on our health.[4]

Yet fast food was, at first, better than many of the alternatives in those days when hunger stalked the streets. And it was an essential part of the new infrastructure of highway travel, for with expanding automobile traffic came rising needs for wholly new systems of supply.

Before we provided food for the rider, however, we had to provide food for the automobiles themselves. The foremost roadside need was for *gasoline*. The past merges so silently into the present that it's easy to forget how much change we've seen in serving that need. That's because the important part of change occurs on a deeply subjective level. Change occurs on the level of textures, tastes, and smells.

I caught a whiff of the 1930s when I ran across another book that brought back the fading smell of gasoline. The title was *Gasoline*. It was a catalog of the SIRM (Società Italiana Ristrutturazione e Manutenzione) Museum in Milan, Italy. The museum celebrates the way we dealt with gasoline in the early days of automobiles. It takes us from the first crude roadside gas pumps before World War I through the 1960s. We see gas cans, logos, and gasoline paraphernalia.[5] And, as I read, a childhood suffused in the smell of gasoline came back to me.

We filled our cars with dime-a-gallon gas and slopped the overflow onto the pavement. We loaded five-gallon cans of the stuff into our trunk in case we ran out between towns. When we sang the song "La Cucaracha" in my high-school Spanish class, we censored it. "The Cockroach cannot travel on, because he has no *marijuana que fumar*" (no marijuana to smoke) became, "The Cockroach cannot travel on because he has no *gasolina par andar*" (no gasoline to use).

We truly did live in a gasoline-drenched world. No one worried about protecting this young boy from its free use. I would bike over to the gas station and buy pints of *white gas*. (That was the name we gave to unleaded gas at the time.) I used it in my model airplane motors, mixed with motor oil so it could lubricate as it burned. Gasoline was omnipresent. We cleaned things with gasoline, we started fires with it. We all inhaled the fumes of our new motorized world, and I had largely forgotten all that until I saw the SIRM book.

A 1946 Ohlsson-23 spark-ignition model airplane motor. The tank on the left held a mixture of roughly four parts white gas and one part heavy motor oil. (Photo by John Lienhard)

The gas pumps of my childhood had big glass containers on top. First, you filled them with the number of gallons you wanted. You began the ritual by *displaying* this new essence. Only then did you let it gurgle down into your tank. (By the way, that's what the French call their gasoline: *l'essence*.)

The SIRM book shows the containers used to carry the stuff around. Those cans proudly display names like Shell, Pennzoil, Conoco, and Standard. Today's pumps hide their essence away. They protect us from sensate exposure—from spills or smells, from fire or cancer. They protect us from ourselves.

And what of the places that sold the new essence? Today's Yellow Pages list those emporiums under the term *Service Stations*, although few offer any service these days. The first public gasoline servers were simply called *filling stations*. They were just curbside hand pumps, and they began appearing in 1907. There was no service in 1907, either. You did your own repairs in those days.

Henry Ford and the great explosion of automobiles after World War I changed all that. Those millions of new cars needed a huge infrastructure of supply, and a new American institution had to come into being: We had to invent the *gas station*.[6]

The gas station started taking shape around 1910. By 1920 it was well-defined. It was a small building with gas pumps in front. It also offered supplies—tires, batteries, and oil. It offered simple services—lube jobs and tire patching. By 1920, America had 15,000 gas stations and only half that number of curbside pumps. By 1930, we had over 100,000 gas stations, and curbside pumps had all but vanished.

Those gas stations became a new American icon. They joined the mimetic architecture movement from the beginning. They made their visual impact in several ways. Some were built like cozy bungalows to welcome weary motorists. Some were futuristic, calling up a modern world of speed and function. Most offered uniformed attendants. And they carried the logos of the vast new corporations now serving up this ubiquitous new fuel—Standard Oil, Cities Service, Conoco, Pure, Sinclair's dinosaur, and Mobil's flying red horse.

Magazine advertising said very little about the new gas stations. Down through the '20s, '30s, and '40s, the various oil companies pressed their *products*, not their *stations*. Those magazines are filled with ads for motor oil, spark plugs, tires, piston rings, and brake shoes—but seldom gasoline itself. The advertisements strove, first and foremost, to groove name recognition into our minds.

Radio advertising sometimes spoke of the way stations would receive motorists. Oil companies told us they were good friends who wanted to serve our needs. "You can trust your car, To the man who wears the star." And, indeed, one often could. However, radio was also primarily concerned with establishing name recognition. My uncle was a distinguished radio journalist in St. Paul: *Brooks Henderson, the Phillips 66 Reporter*. Before we had network news, Phillips 66 had become synonymous with radio news in St. Paul.

The oil companies let *architecture* carry the major burden of calling us into their gas stations, and it did so very effectively. Anyone raised in that era had a distinct sense of each gas station's personality. And the stations were everywhere—they had fewer pumps than you find today, but more locations. By 1970, America had over two hundred thousand gas stations.

Then things began changing. By 1990, half those stations were gone; now they hardly exist. Today we go to a *service station* for gasoline, candy, soda, and a car wash, although we seldom expect to find even such elementary services as an oil change or a new battery. Today, batteries, oil, and tires last five times as long as they once did, and as cars grow more complex, much of their upkeep moves back out of our hands.

The gas station is a cultural icon that has all but vanished, and most of us are only vaguely aware of its death. Indeed, it is hard to reclaim its visual presence. Helms and Flohe's book, *Roadside Memories: A Collection of Vintage Gas Station Photographs,* includes relatively few images before 1950.[7] We registered those small buildings as we whizzed by, but we were hardly about to turn our cameras upon them. As far as we were concerned, they existed only for the few seconds it took us to drive by— at least until our tanks ran dry.

Now they are gone, and so too is a whole way of life: White Castles, drive-in movies, the Teepee Motel . . . Driving up to the gas pump, we no longer pass over that rubber tube and sound a bell, summoning a liveried attendant to come out and treat us like visiting royalty.

Now we buy gasoline from a pump that looks like a computer interface: "Insert your credit card here." "Make sure the seal is tight before you add fuel." Instead of celebrating the liquid that we all take for granted, today's gas pumps hide it away. We no longer get into the car smelling of fuel. We live longer. We live safer. We live far away from those rich, unsafe old celebrations of our new mobility.

The highways themselves were the last part of the evolving infrastructure to catch up with our speeding automobiles. I worked for the Bureau of Public Roads in the summer of 1949, laying out a gravel forest-access road in the Cascade Mountains, practically in Canada. I was, at the time, quite oblivious to another activity going on within the bureau. Behind small projects like this, being carried out all over America, the bureau was hatching a truly revolutionary plan back in Washington. It was the 40,000-mile Interstate Highway System.

The Eisenhower administration finally backed the system in 1956. It was to be finished by 1971. As it neared completion, author Helen Leavitt published a scathing indictment of the project called: *Superhighway—Super Hoax.*[8] I had no recollection of the book until I found it in the New York Public Library's list of 150 major books of the twentieth century.[9]

The title accurately represented the attack leveled by the book itself, and I was left wondering: Was Leavitt the greatest Luddite of this century, or did she have a case? Who would condemn this magnificent system of superhighways today? It is a primary and accepted part of our lives.

The system had surely been helped along by Eisenhower's recollections of World War II. Germany's autobahn supply system had been a thorn in his side. He wanted as much for America. Indeed, the project was named, "The National System of Interstate and Defense Highways."

So what were Leavitt's objections? Cost was certainly one. Once Congress began considering the undertaking, oil, automobile, and concrete lobbyists flocked to Washington to support the $50 billion project. A new car cost $2,000 back then. Once you had it, you were committed to spending three times that initial cost on gasoline, insurance, maintenance, and highway-related taxes.

The highways also ate real estate without mercy. Homes, buildings, and farms were plowed under in draconian seizures of private property. Today, that damage is largely forgotten, and we've long since built the costs into our budgets. But another Leavitt objection is still with us.

Our whole way of life was being built around the automobile to the exclusion of the various urban rail systems. By 1980, half the surface area of Los Angeles was cast in highway concrete. The neglected national rail system was falling apart, and so were city transit systems. Hydrocarbon emissions had become a major health threat. So, was Leavitt a Luddite or visionary?

The fact is that, when I worked on roads in 1949, our marriage to the automobile had long since been consummated. Building the Interstate System was no point of departure at all. We had made our decision decades earlier. Besides, as we tore across America at 80 miles an hour, joyously guzzling cheap gasoline, only a crank could have seen the dark side of those excesses.

Even now we have to squint our eyes to see the wreckage created by a total automobile economy. Leavitt showed pictures of comfortable new buses and railroad coaches, of the high-speed trains that Japan and Germany went on to make a reality. Yet she surely *was* a Luddite in 1970, for *Modern* had shaped an automobile economy long before the highway system struggled to catch up with it. The visceral force of *Modern* had, long ago, run away from the logic of even the wisest Monday-morning quarterbacks.

Consider, for example, the way in which the elements of *style* flowed from the automobile into daily life. I invite you to form the following image in your mind—one part of the conversation between automobiles and the side of the road: Imagine yourself walking on the shoulder of a road. Think about what happens in the pit of your stomach when a car passes within a few feet of you, moving with the speed of a gale wind. Like the carcinogenic presence of spilled gasoline, that sensate experience was once omnipresent. But few of us know it today.

I knew it well from hitchhiking along two-lane highways. A ton of steel, passing a yard away at 70 miles per hour, creates a profoundly impressive blast of air and sound. Closing my eyes and recalling it now reminds me just how little sobriety or safety was to be found in any of our modes of transportation when they first emerged.

Early steamboats and railways were wretchedly dangerous. Yet, if you read Mark Twain, or if you sing "Casey Jones," you know those dangers were the subject of celebration more than lamentation at the time. (More about that in chapters 10 and 11, where we look at flight.)

And one need only count auto fatalities per passenger mile to know how safety has steadily improved. Eleven times as many people died per passenger mile in 1925 as in 1997. When I was young, most high school graduating classes of any size had been reduced by a few automobile deaths.

So we engaged the wind—either beside the road or from within the car. Turn and stop signaling was all done by hand from an open window. As a bored passenger, I would roll down my window and fly my hand in the wind, shaping it like an airfoil and learning how the forces of the wind acted upon it. The only care I took was to stay away from the left back seat, where my play could be confused with the driver's hand signals. That sort of thing was more dangerous than anyone ever stopped to consider, of course. However, it left us with a very clear picture of how much drag air imposed at highway speeds.

Thus the automobile helped to form the next important shift in design style. Speaking in very simple and broad terms, a sequence of three primary design styles was associated with *Modern*, all driven in one way or another by the new technologies. First *art nouveau*, then *art deco*, and, at last, *Modern* brought us to *streamlining*.

Art nouveau, with its twining vines and sensuous curvatures, had lain at the origin of the arts and crafts movement. John Ruskin, with his loathing of nineteenth-century technology and his craving to find a way back to elemental nature, was the influence behind it. Ruskin's disciple, William Morris, began carrying those ideals to a larger public during the 1860s.

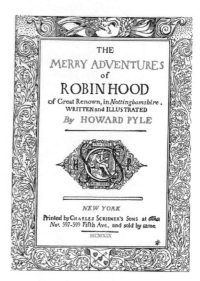

Morris dabbled in crafts of all sorts and did more than anyone to set the art nouveau style. In his book *A Guide to Art Nouveau Style*, William Hardy remarks that "Art Nouveau developed in the late 1880s and was at its creative height in the subsequent decade. By 1905 it had declined into being a much diluted ingredient in commercial design, soon to be replaced by an aesthetic felt to be more in keeping with the new century."[10] I say more about that in chapter 13, but it was in that period of "dilution" that art nouveau became a kind of retro battleground against the floodtide of *Modern*. Tiffany lamps and Klimt paintings expressed, very strongly, our yearning to retain the grace and form of nature as we hurtled ahead.

Would-be art nouveau ornamentation is shot through this 1919 edition of *Robin Hood*.

I talk in chapter 6 about how art nouveau gave way to the clean, simple linearity (and verticality) of art deco. By the time Wanamaker's department

The 480-foot Bush Terminal Building in midtown Manhattan was begun in 1916. This fine art deco representation of the structure appeared just as it was finished, from the Sept. 1921 *Century Magazine*, p. 714.

store exhibited its pure art deco images of the titan city (chapter 7), this new design medium had been rising with the new skyscrapers for at least a decade. Art deco made a superb vehicle for shaping the functionality of the new skyscrapers into a new and powerful subjective artistic impulse.

Now the fluidity of horizontal movement had to replace that static verticality. We needed a new design metaphor that could wed itself to the wind. Many people had been studying the way moving objects were affected by the air through which they move. Gustave Eiffel did early work with wind tunnels after he erected his tower. The Wright brothers built their own wind tunnel, which was a crucial part of their development of an airplane.

In the years just before World War I, the German scientist Ludwig Prandtl formed the Kaiser Wilhelm Institute for Flow Research at the University of Göttingen.[11] The studies of the mechanics of lift and drag that he and his staff did there formed the basis for the new science of aerodynamics.

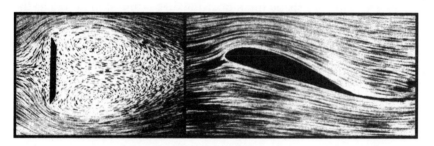

These images have been adapted from early flow-visualization photos, originally made by Prandtl at the University of Göttingen. They show how the shape and orientation of a body affect the aerodynamic forces that act upon it.

Airplanes and race cars had already begun reflecting a new awareness of aerodynamics. Goggled pilots and race-car drivers with their faces in the wind could not help but understand the forces they were dealing with. Now the work of Prandtl and others began informing airplane designers in very specific ways that they might optimize the forces on an airplane.

But we need to remember that airplane speeds did not catch up with those of automobiles until around 1920. And streamlining did not establish itself as a new design style until Chrysler shocked the public with his 1937 Airflow automobile design. The Airflow was, to many eyes, an ugly duckling, but the idea was now afoot and would not be denied. Indeed, Chrysler was far from first. One cannot mention the new streamline-based design movement without including Raymond Loewy.[12]

Loewy was born in France in 1893. When World War I interrupted his design studies, he joined the artillery. He won the Croix de Guerre in combat, but seemed far more focused on the challenge of redesigning his uniform than on war. Afterward he joined his brother, a doctor, who had been called to New York to work with poison-gas victims.

Loewy was stunned by New York. The immensity, intensity, and lack of any design order, offered him a clean slate. At first he did freelance product design for Wanamaker's store and Condé Nast. Then he received a commission to redesign a mimeograph machine. When he was done, an item that had looked like industrial machinery had become art deco office furniture.

After that, all kinds of manufacturers went to Loewy for product design. He gave us the vertical lines of refrigerators we use today. His streamlined diesel locomotive design swept the American imagination. He designed the 1934 *Hupmobile*, which was the less-well-known, streamlined precursor of Chrysler's *Airflow*.

He gave streamline motifs to everything from clothing to pencil sharpeners. He helped to shape the modern luxury cruise ship as well as the city bus. We even find lingering elements of Loewy's streamlines in today's farm tractors. Perhaps Loewy is best known for one of the late icons of *Modern*, his famous (and even beloved) 1950 Studebaker design, with its futuristic streamlined headlamps.

In many minds, the very word *postmodern* summons up the world that can no longer be identified with Loewy's designs. By 1960, he had defined America with such clarity and finality that we've spent a half-century looking for something strong enough to follow that wild design impetus. Late in life Loewy wrote that he had come here sixty years before and "immersed myself in the American Vortex." That's a wonderful line, but the vortex itself was in great measure, a thing of Loewy's own making.

The new metaphor of streamlining ignited the public consciousness like a spark in tinder. It provided the expression of *Modern* that we'd all been seeking. Now, at last, we could leave the old design virtues that had been exerting some residual restraining force upon our breakneck forward motion; we could cut the last cords still binding us to the nineteenth century.

Yet this culmination of *Modern* was also a last hurrah. A few years later, as airplanes finally exceeded the speed of sound, carmakers began putting large rocket-like tailfins on some of the heaviest automobiles yet offered to the general public. Those tailfins served no real aerodynamic function and, as we looked at them, we realized were seeing a *reductio ad absurdum*. They practically shouted, "Okay, just try to top this!"

(Photo by John Lienhard)

Of course, we could not top that. We needed a new direction. More on that in chapter 16, but now let us turn around and drive back into the world of the early twentieth century. The time has come to visit what could well the defining technology of *Modern*—one of its purest expressions, the technology of *flight*. We look next at the way in which flying became a magnet to the people who most needed means for improving the quality of their lives, a century ago.

10

The Back Door into the Sky

I suggest in chapter 3 that 1901 might serve as our *annus mirabilis,* the wondrous year in which *Modern* found its place upon the public stage. That was also the year the Wright brothers first flew. They were still two years from *powered* flight, but they made several successful glider flights in 1901, and they validated their new system for controlling an airplane in flight.[1]

The Wright brothers were not the first to fly. Gliding had been done over a thousand years earlier. Humans had also flown in kites, balloons, and even, to a limited extent, powered, heavier-than-air machines.[2] But the Wrights finally did the three things needed for a complete success: they put the apparatus of modern science fully to bear upon the whole

A Wright brothers' glider flight from Kill Devil Hill near Kitty Hawk, from the 1907 *Century Magazine.*

problem of flight, they made controllable and repeatable flights, and they brought an airplane all the way to market.

Consequently, the Wrights created a watershed in the age-old struggle to fly. Flight would, henceforth, be a part of the rush of newness that was *Modern*. Looking back upon what was written about flight, one might well conclude that, before the Wrights, anyone who had been involved in trying to fly was, ipso facto, a crank, and that afterward they were builders of the brave new world of the twentieth century.

However, *anni mirabiles* are always oversimplifications. The transitions that they represent are never completed in a mere 365 days. If early pioneers of flight were often marginalized before the Wrights, that marginalization lingered after they succeeded. Furthermore, intimations of legitimacy were to be found here and there throughout the nineteenth century. The story of Alphonse Penaud illustrates the mechanics of this complex process very accurately.

Penaud was born in 1850 in Paris.[3] He had planned to join the navy when he grew up, but then he suffered a disabling illness that ended his plan. As a late teenager, he turned instead to the problem of inventing a flying machine. Penaud was no idle dreamer. He worked methodically, carefully thinking through all the issues of stability and propulsion.

In 1871, he perfected a rubber-band-driven model airplane, which he called a *Planaphore*. It had a wing, a tail, and a propeller in the rear—a design remarkably close to the models that kids like me built 70 years later. It was roughly the same configuration we were using for most airplanes when World War I began. He flew his model publicly in the Tuileries Gardens, and, during the next ten years, he kept on experimenting.

Penaud's Planaphore, as illustrated in the 1897 *Encyclopaedia Britannica*.

In 1875, the French Academy of Sciences awarded Penaud a prize for his brilliant thinking. Still, he meant to fly, not just think. He went to work with a mechanic named Paul Gauchot, and in 1876 they patented a full-scale airplane. For the next four years, he looked for the means to build and fly the machine.

Long before, when Penaud was just a two-year-old child, another French inventor, Henri Giffard, had made the first successful dirigible flight over Paris. Giffard was, by the 1870s, France's leading experimental aviator. So Penaud turned to him for support in getting a heavier-than-air craft built. Giffard gave him a cold shoulder.

Penaud did not have the emotional resources to handle that defeat. He built a small wooden casket and put all his designs into it. He delivered it to Giffard's house and then, at the age of only 30, went home and committed suicide. What he didn't realize was that Giffard was also depressive. Two years later, Giffard himself committed suicide.

The pioneering work of both Penaud and Giffard had a huge impact. Read any book on the history of flight and Penaud's intellectual work is there as a recurring footnote. Read any history of dirigibles and it will begin with Giffard.

Penaud made one of his most enduring contributions when he created his rubber-band-powered helicopter. It was sold as a children's toy, and as we read in chapter 1, Bishop Milton Wright touched history by bringing one home to his boys, Orville and Wilbur, in 1874—four years before Penaud's suicide.

Both Penaud and Giffard cracked under the stress of their marginal position in the world of invention. Trying to fly usually carried with it some stigma of madness. Airplane inventors were often among those who, for one reason or another, felt disenfranchised. Remember (from chapter 5) quiet Charles Dellschau, who lived within his own head and dreamed of mysterious aircraft over the forests of Sonoma County, California.

The mirror image of that state of affairs was the way in which early flight also provided a new portal *out of* marginalization for those whom society had marginalized. Once flight began finding its way into legitimacy, some remarkable social realignment occurred. Flight was, for a season, a significant agent of social change. Women and black fliers, in particular, emerged in this new arena, where bars had not yet gone up. They found, in the new flying machines, a means of *escape.*

The theme of flight as an escape from society's shadows goes back 3,400 years. Our primary ancient legend of flight tells how Daedalus, jailed on the island of Crete, fashioned wings of feathers and wax so he and his son Icarus could fly to sanctuary in Sicily. When Icarus flew too near the sun, the wax melted, the feathers blew away, and he fell to his death. Daedalus made the flight safely all the way to Sicily.

Daedalus's flight, Icarus's fall, and the dream of escape to a better life have touched our minds ever since. Most ancient legends have some basis in history, and I would be very surprised if there were not some historical Daedalus who actually did experiment with flight.

As we go forward in time, we trace a mutation of the stuff of legend into reports that we call *historical.* The ninth-century Spanish Moor Ibn Firnas built wings and, like Daedalus, covered them with feathers. Being a flesh-and-blood historical figure (living on the fringes of history), he crashed and hurt his back. About that same time, Vikings began telling a story that echoed both Daedalus and Firnas yet included a new insight. Like Daedalus, their hero Wayland was also a prisoner, and he too made feathered wings to escape. When his brother Egil tested the wings, he crashed because he had failed to launch himself into the wind.

These insights converge in a story told by the twelfth-century English historian William of Malmesbury. He writes about an Anglo-Saxon named Eilmer. Eilmer was not a prisoner, but a monk at Wiltshire Abbey. Like Daedelus, Firnas, and Wayland, he also launched himself from a high place—a tower. Like Ibn Firnas, he failed to equip himself with a tail for lateral stability; however, he did dive into the wind. He sailed over a hundred yards but broke both legs when he crash-landed. All the same, his exploit also contributed to the mounting corpus of historical (that is, scientific) experience.[4]

Since all of these pioneers of flight from the murky past were male, I stopped a classical scholar one day and asked if his studies had ever revealed a *woman* in flight. He looked up at the trees and thought. Then he remembered how, in *Phaedra*, Plato created the following image of the human lot: At first all men and women alike were endowed with feathered wings and flight. Then *Vice* gradually pulled us down until our wings were no longer feathered and we could no longer fly. (My friend also recalled that Plato had called Love "the feathers of our souls.")

But the creation and use of flying *machines* has been reported as an all-male activity through most of history. Even the angels were always portrayed as male before the nineteenth century. Their gender changed just after people had begun creating flying machines—just when flight began seriously mutating from a dream into palpable reality.

The Montgolfier brothers created the first manned hot air balloon, which made its first ascent on November 21, 1783.[5] Eleven days later, Alexandre Charles tested a manned *hydrogen*-filled balloon. When one early hydrogen balloonist, Jean Pierre Blanchard, finally suffered a fall and died, his wife Sophie continued in his barnstorming business. She had taken up flying in 1805. Sophie Blanchard was the first woman we know to have flown on her own, and she made 59 ascents before she too died. She accidentally set her hydrogen on fire, and, as it burned off, the balloon fell to the Paris rooftops below.

Even before Sophie Blanchard, women had been taking part in another kind of aerial display, which is powerfully evocative of the Daedalus story. In 1797, André Garnerin became the first person to parachute from a balloon. Like Daedalus's wings, the inherent purpose of a parachute is to

Model of the Montgolfier balloon in the National Air and Space Museum. (Photo by John Lienhard)

provide a safe avenue of escape. Garnerin talked his wife into making one jump, but it was his niece who saw the parachute as a path into a new and richer life. She took up exhibition parachuting for a living, and women continued to make exhibition jumps throughout the following century.

That brings us to Buffalo Bill Cody, who, the same year the Wright brothers flew, took his Wild West show to London. One of Cody's acts involved balancing a plaster egg on his wife's head and then shooting it off with a pistol while he was blindfolded. At one of his shows, Cody aimed just a hair too low, and the bullet grazed his wife's scalp. A 16-year-old girl, Dolly Shepherd, came out of the audience and volunteered to take the place of Cody's wife in the act.

To thank her, Cody took her to an aeronaut's workshop. There, Shepherd found work as a parachutist and soon became the star of the troupe. It was heady and dangerous work for a young girl. She saw her first fatality when one of the balloons used by the troupe lost its buoyancy, landed on a factory roof, and dragged a girl over the edge to her death—exactly the death that Sophie Blanchard had suffered 84 years earlier. Shepherd herself was almost killed several times.

In a typical jump, two girls got ready. They opened a vent hole in their balloon so it would start down. Then they jumped. One day, after Shepherd had vented the balloon, she found the other girl's ripcord was jammed. The two had no choice but to jump together from 11,000 feet using Shepherd's small parachute.

Shepherd was badly shaken up, and, while she was getting her strength back, her mother secretly jumped in her place. Shepherd jumped again, but she began seeing the face of Death in all of this and gave it up. Two years later she took up safer work; she joined the war in France as a driver-mechanic.

When Dolly Shepherd died in 1983, she was 96—old enough to have seen people on the moon and a rocket circling Saturn. She had lived to see women in rockets that were just as new and primitive as the parachutes she had used 80 years before.

So this seems to have been more than just a tawdry business of pushing women over the edge as a source of cheap excitement. Flight also proved to be the means by which many women, tied down in the social climate of a century ago, managed to slip the all-too-surly bonds of Earth.

Flight had other dimensions as well: of style, fashion, and theater. As an example, I give you the Contessa Grace di Campello Della Spina.[6] In 1907 this aristocratic lady wrote an article that begins with the words, "Sport of the Gods!" Queen Margherita of Italy had founded the Roman Aero Club in 1904, and the Contessa had become an ardent member. Her sport of the gods was ballooning.

First she had flown on a tether rope. Now, on the occasion of her honeymoon, she prepares for her second ascent. She, her new husband, and a pilot prepare to fly across the Apennine Mountains to the other side of Italy. They take off at 9:30 in the evening, and the Contessa rhapsodizes, "A full-orbed moon was just appearing in a mist of golden glory above the Alban Hills . . ." She tells how to pack for the trip: "Bring the simplest and most practical tailor-made suit of light wool. Take an umbrella. The sun bites hard in the South." She advises balloonists to choose their food carefully: "It should be light and easily eaten. French Prunes, raisins and chocolate—cold tea, mineral water, and well-baked bread. Bring a light volume of your favorite author, and no alcohol save a small flask of brandy in the case of faintness."

As they rise into the night sky, their view is lovely. They see their reflection in a pillar of cloud by moonlight. At length the Contessa describes dawn: "On the wings of the stiffening morning breeze we raced along in joyous flight like happy swallows."

Their balloon has a rip valve so they can let gas out and land, but it's better, she tells us, to throw out a rope so the peasants can tow you in. They do that, and the trip ends in a cornfield. She says, "The peasants, seeing us descending, ran from all sides to help pack the balloon, and claim damages for the corn." She sits on the passenger car and chats with the local women. She observes that, "They are most primitive and full of rustic curiosity."

(*Left*) Artist's conception of the Contessa's description
and (*right*) The Contessa emerges from the balloon's gondola.
(From the May 1907 *Century Magazine*)

The next day the Contessa and her husband catch a train back to Rome. It is hot and uncomfortable, and the trip takes them 24 hours on the ground. She is back to the hard earth, and the honeymoon has ended. A few years later, those balloons would become a new weapon in the most terrible war that had yet to afflict our planet. When the smoke cleared, we were vigorously dismantling the Contessa's royal class.

In the meantime, with all that money to lavish on play, the Contessa really had been a valid ally of human progress. Her flight had been all about pleasure and adventure-seeking, but name a more powerful driver of technological change than pleasure, if you can. I cannot.

There was nothing disenfranchised about the Contessa Grace di Campello Della Spina in 1907, of course. Nor was one Mrs. Hart Berg outside society's inner circles a year later, when Wilbur Wright appeared in France to demonstrate the new Wright airplane. The American industrialist Hart Berg saw France as a ripe market for the Wright airplanes and was helping Wilbur to sell them. It was a fine idea. France, in historian Fred Howard's words, "clasped this son of Dayton to its bosom as it had no other American since Benjamin Franklin."[7]

Wilbur set up camp near Le Mans, where he made 120 flights. In one of those, he carried Hart Berg as a passenger. He then took Berg's wife up for a flight, an act that turned out to be much more significant than Wilbur realized at the time.[8]

For one thing, Mrs. Berg became the first American woman to fly in an airplane. But who could have predicted the other outcome? Before the flight, Mrs. Berg tied a rope about the bottom of her skirt to keep it from blowing in the wind. A French fashion designer saw her mincing steps as she left the airplane afterward and was fascinated by the possibilities. Mrs. Berg had, on the spot, inspired what would, within months, become the latest fashion craze, the *hobble skirt*.

This particular woman aloft had unwittingly found, and tied into, that metaphorical substrate underlying any of our technologies. Flying was about to move from the margins into high fashion. And, as women in the air crossed the threshold into fashion, fashion and suffrage would now become interwoven.

Women from every walk of life flirted with fashion on the surface as they became deadly serious about flying. Two years later, Blanche Scott became the first American woman to pilot a plane, and her story echoes the shift-

Wilbur Wright and Mrs. Hart Berg take to the air, from a 1910 *Cosmopolitan* magazine.

ing drive toward suffrage. She managed to get into Glen Curtiss's new flying school, but since Curtiss did not like the idea of women flying, he blocked the throttle on her plane so it would only taxi. Scott somehow managed to override the block and get 40 feet into the air.

Two weeks after Scott's brief flight, a woman named Bessica Raiche soloed quite intentionally, but she soon gave up flying to become a doctor. Julia Clark was the first woman to die in a crash. She learned to fly in 1911 and died practicing for an exhibition two years later.[9]

Many other women took to the air, but one in particular catches our imagination, as well as the mood of the times. Harriet Quimby brought fashion, theater, and stunning independence to the airplane all at once.[10] Indeed, one cannot tell Quimby's story without commenting upon her commanding beauty and style. Beauty, however, is a created quality and, in her case, it cannot be brushed off as an accident of nature.

Quimby was born on a Michigan farm in 1875. The Quimby family moved to San Francisco during Harriet's early teens. She wanted to become an actress and is, in fact, listed as one in the 1900 census. If she ever did any theater, however, there seems to be no record of it. Instead, she began writing for various magazines and, in 1903, went to New York and found work with *Leslie's Illustrated Weekly*.

She became a full-time photojournalist whose writing leaned toward excitement—travel, theater, racecars, flight ... In 1906, after a ride on an automobile racetrack, she bought her own car. Four years later, in late October of 1910, she covered a very early international aviation tournament. This time she came away determined to become a flier.

One pilot in the tournament was John Moisant, who, with his brother Albert, was also manufacturing a monoplane of his own design. Harriet Quimby was a close friend to John, Albert, and their sister Matilde, so she asked John to teach her to fly. John agreed, and a number of events quickly unfolded.

First, John was killed two months later in an air show in New Orleans. The Moisant School of Aviation nevertheless opened on Long Island only three months after his death. Both Harriet and Matilde enrolled. On August 1, 1911, Harriet Quimby became the first woman licensed to fly in America (second in the world). Matilde soon followed.

Harriet created a trademark purple flying costume for herself. It consisted of a satin jacket with a soft cowl around her head, high laced boots, and satin riding pants. Tall and elegant, she was unmistakable at a distance, and close-up photos show a dark-eyed ethereal face looking out from the shadows of that cowl. One wonders if she might have flown in from the recesses of pure imagination.[11]

Quimby was an instant hit on the barnstorming circuit, and her articles about flying carried her popularity to a larger public. That spring

Harriet Quimby getting into her Moisant Monoplane, from *The American Review of Reviews*, 1911.

she went to England with the purpose of buying a Bleriot airplane. She borrowed one in Dover, and, early on the morning of April 16, 1912, she took off to become the first woman to fly the English Channel. But fate got in her way. Hours earlier the world had learned about the sinking of the *Titanic*, and that wiped Quimby's feat from the headlines.

So she returned to America and more barnstorming. On July 1, 1912, she was paid handsomely to do an air show near Boston. In front of everyone, her plane lurched, throwing her passenger to his death. Quimby struggled to regain control; then she too was thrown to her death. The empty airplane glided in, landed, and flipped over in the mud. Harriet Quimby had flown for less than a year, but she wrote history while she did. While she was at it, she surely became the actress she once aspired to be.

Women amassed firsts and set records during those years just before World War I. Naturally, as it became clear that airplanes would play a role in the war, women wanted a piece of that action as well. Then the wall went up; the military strongly opposed involving women in war.

The woman who got closest to aerial combat was Katherine Stinson, older sister of the noted airplane builder Eddie Stinson. When I was a kid building model airplanes, one of my favorites was the Stinson Reliant, a fine airworthy machine that seated four people. My friends and I had heard of Eddie Stinson, but the names of his siblings, Katherine, Marjorie, and Jack, were completely unfamiliar.

Still flying, this Stinson Reliant is used by Tillamook Air Tours in Tillamook, Oregon. (Photo by John Lienhard)

Katherine, born in 1891, was the serious older sister.[12] In 1910, seven years after the Wrights flew, she was focused on finding a way to study music in Europe. To do so would require $1,000 that she did not have. Then she read about the new stunt fliers and realized that the best barnstormers might pick up $1,000 in one good day, if they could stay alive.

In 1912 she located flying pioneer Max Lillie, who had a flying school in Chicago. Would he teach her to fly? Lillie looked at Katherine's five-foot, hundred-pound frame and said, "Not a chance!" She finally persuaded him to at least take her up for a ride. When he did, he was struck by her utter calm and self-possession. She may have been small, but she was a natural.

Her mother supported her ambition, but Katherine had to wear her father down. He finally paid half of her flying-school tuition, and she (in a move reminiscent of O. Henry's story "Gift of the Magi") sold her piano to get the rest. She became the fourth American woman with a pilot's license. Next, she claimed to be only 16 (instead of 21) and took up exhibition flying. She billed herself as "the Flying Schoolgirl."

During an exhibition in Helena, Montana, the local post office obtained authorization from the U.S. Post Office for an airmail route between that Helena fairgrounds and downtown Helena. That route, of course, was itself only a stunt. But it nevertheless made Katherine Stinson America's first woman airmail pilot in September 1913. She carried a total of 1,333 letters and postcards.

She kept up the pace. She was the first woman to fly a loop. She flew exhibitions in Japan and China. In 1914, her sister Marjorie became the ninth woman pilot licensed in America. Marjorie learned to fly at the

Wright brothers' school (but only after her mother had wired Orville her permission). Eddie Stinson got his license in 1915.

That same year, the family moved to San Antonio, Texas, and formed the Stinson Aviation School. Katherine, Marjorie, and their mother, Emma, ran it. Eddie was the mechanic. The youngest son Jack (then only 15) would later learn to fly and would form a school of his own. When America entered World War I, the military banned civilian flying, and the Stinson school closed down. By then, Marjorie was training Canadian pilots, who called her "the Flying Schoolmarm." That was in stark contrast to her sister's Flying Schoolgirl persona. Women's relationship to flying had seriously shifted.

When Katherine tried to enlist as a pilot and was rebuffed, she went to the Red Cross. They sent her on a flying trip around the country in a Curtiss Jenny to raise funds. (She set new distance records as she did.) In 1918, she finally got permission to go to France as an ambulance driver and as an occasional flier for the Red Cross.

Katherine Stinson's younger sister Marjorie was also sworn in as an airmail pilot in 1915. The mother of the Stinson family, Emma, is shown second from the left. The airplane appears to be one of the Wrights' designs. (Photo courtesy of the Smithsonian Institution archives)

After the war, Katherine Stinson went back to flying airmail, but she came down with tuberculosis in 1920. After a long recovery, she married a former World War I pilot. They both did a little flying, but, in 1930, they decided to quit altogether. She did drafting for the army and then took up architecture. She won prizes for her designs, and she lived to the age of 86.

Eddie Stinson, who designed some of the great light airplanes between his bouts with alcoholism, is the Stinson my generation remembers. But Katherine, daring and methodical Katherine, was the driving engine behind it all. A saga was begun when she sold her piano in a Byzantine plan to study music in Europe! She got to Europe all right, and she did shape a rich life by doing so, but it was a life very different from the one she had set out to create when she was a teenager.

While women like Shepherd, Quimby, and the Stinsons were writing their aerial declaration of independence, black Americans were also looking to the sky for a door into freedom and acceptance. The first black flyer in any nation was Eugene Bullard.

Gene Bullard was born in Georgia in 1894. His father was a black laborer with some education. His mother was a Creek Indian who died when he was only six. Author Jamie Cockfield tells how Bullard's father told him stories from books he had read.[13] Bullard grew up knowing there was a country, far away, called France—a place where he had heard that black people were treated like human beings.

Photo of Eugene Bullard as displayed at the National Air and Space Museum, Washington, D.C.

That lay upon his mind. While he was still a boy, he ran away from home and worked his way across America toward France. He stowed away on a transatlantic freighter, was caught, and was put ashore in Scotland. From there, Bullard worked his way south through England while he took up prizefighting. He had a talent for boxing and was given fights in England, North Africa, and Germany. Then, in 1913, he was sent to Paris for a bout.

Bullard had made it to France at last, and Paris was all he had dreamt it might be. So he stayed. When war began in August, 1914, he joined the French Foreign Legion. The day after his first terrible action in the Battle of Artois, only 1,700 of 4,000 men in his outfit answered roll call. He fought in Champagne and finally at Verdun, where a German shell blew a hole in his leg. A *Saturday Evening Post* writer caught up with Bullard in the hospital and wrote that he was a "[confident] black Hercules . . . not at all like the Negro we knew at home." Bullard's wound left him unfit for the infantry but, if he walked with a limp, that was of little concern to the new French flying service. In 1916 they took him into flying school, where he trained and qualified. He left for the front and brought down a German airplane in his first dogfight. Then he sputtered back to the airstrip with 78 bullet holes in his airplane. He shot down a second plane over Verdun.

After the war, Bullard opened a nightclub in Paris and stayed until the Germans entered the city in 1940. At that point he had to flee, since he was a sure candidate for the concentration camps. An angry Gene Bullard found his old infantry unit and fought with them briefly—until Germany defeated France. Then he had to flee again, this time by way of Spain, back to the America he had left so long before.

Bullard's last job was operating an elevator in the RCA Building. Dave Garroway spotted him there and gave him a guest spot on the *Today* show. This hero, with his rows of French medals, and who had been embraced by General de Gaulle, finally received his 15 minutes of fame before he returned to the elevator.

Bullard's personal insignia had been a red heart with a knife through it, and the words, "Tout le sang qui coule est rouge!" (All blood runs red!) His last words carried that same power of double meaning. Lying in New York's Metropolitan Hospital in 1971, he suddenly seized a friend's hand and gasped, "It's beautiful over there." Then he died.

The second black American to fly was a woman, Bessie Coleman, and the theme of Daedalus escaping his prison is as clear in her life as it was in Bullard's.[14] She was born, the thirteenth child in her family, in 1892. Early in her life the family moved to Waxahachie, Texas, where they lived on the edge of poverty. They picked cotton in a region where lynching was not uncommon, and the Ku Klux Klan was highly active.

Bessie was smart and determined. She learned to read, and she checked out books from a local library. She read to her family at night: the Bible, *Uncle Tom's Cabin*, books about Booker T. Washington and Harriet Tubman. She finished the eighth grade and then studied at the Colored Agricultural and Normal University in Langston, Oklahoma. Her money ran out after a year there. She moved back to Waxahachie, where she did laundry and dreamt of a larger life.

Bessie Coleman moved to Chicago in 1915 and spent World War I working as a manicurist while she read about the new heroes of flight in France. She decided that she herself would fly. Since no American flying school would have her, she learned French in night school, cultivated contacts among the wealthy, saved money, and sailed for France in 1921.

She managed to enroll in the Coudron Brothers' School of Aviation, where she trained in the Nieuport Type-82—a slightly more advanced version of the airplane my father had so loved to fly when he was in France two years before her. Like him, she wrote about the smells, sounds, and sensations of early flight. She graduated in 1922 and returned to a very different America.

She had been licensed to fly by the best. In the 1920s she joined the community of black intellectuals in Harlem, where she hatched a plan to set up her own flying school. The next year she traveled back to Europe to get advanced training in aerobatics and drum up support for the idea. Anthony Fokker entertained her in Holland. He showed her his airplane factories and vowed his support. Former German pilots entertained her in the Berlin of the Weimar Republic.

Back in America, she barnstormed to raise money for the project, and now her beauty and pizzazz made her a favorite of the white press. She suffered a nasty crash in California in 1923, and her exhibition career went on hold for two years.

She returned to flying and to Texas in 1925. She did air shows in Houston and Dallas. When she went back to Waxahachie for a show, she found that the gates and bleachers were to be segregated. Coleman drew a line in the sand: There would be no show unless black and white entered by the same gate. The management finally agreed. (Once inside, of course, black and white would sit in segregated bleachers.)

Coleman had, by this time, added parachute jumping to her repertoire. In 1926, she was scouting a parachute jump site in Jacksonville while her mechanic flew the plane. They didn't realize that a loose wrench had been left in the cockpit. It slipped into the control gears of her Curtiss Jenny and jammed them. They went into a tailspin. Coleman was thrown out and the mechanic crashed with the plane. Both of them died.

Bessie Coleman was only 34, but she had left behind, hanging in the clear air above her, a remarkable legacy. It was a legacy that touched another young black American in particular. He was William J. Powell, born in Henderson, Kentucky, in 1899 and raised in Chicago.[15]

Powell was a top student with a talent for music as well. He interrupted his studies at the University of Illinois to serve in World War I as an infantry lieutenant. After being badly wounded in a gas attack, he went back to Illinois to finish his electrical engineering degree. All the while, all around him, were the terrible informal segregation of the North, and the strict Jim Crow rules of the South. But Powell believed explicitly that the Negro could repair in the sky what had been broken upon Earth.

In 1934, Powell wrote a thinly fictionalized autobiography, *Black Wings*.[16] It tells how he visited Le Bourget Airfield soon after Lindbergh had landed there, how he took his first airplane ride, and how deeply moving it was. We learn how he was rejected by a flying school and by the Army Air Corps and how he was finally accepted into a Los Angeles flying school in 1928.

By 1932 he was licensed, not just as a pilot, but as a navigator and as an aeronautical engineer as well. In *Black Wings*, he writes, "I do not ally myself with [the] Negro who begs a White man for his job. I ally myself with that . . . young progressive Negro who believes [he] has the brain, the ability, to carve out his own destiny." Powell meant to fly around Jim Crow, and he concluded that black Americans could take hold of the embryonic flight industry to build their own economic independence.

He founded a Bessie Coleman Aero Club and organized an all-Negro air show for the club in Los Angeles in 1931. It drew 15,000 visitors. Powell built his own flying school and shop. He was no longer an aerial showman or a dramatic public figure. Indeed, he was death on sham of any kind, and his book sparkles with down-to-earth technical detail.

An old photo shows Joe Lewis in Powell's workshop, giving his good name to the cause. But Powell gave that cause true substance. He gave it his belief system and his well-honed bourgeois work ethic—the most inexorable force in the world.

He died, still young, in 1942, probably from the after-effects of World War I poison gas. Jim Crow outlived him, but Powell had seen a few of the early fruits of his work. The Air Corps formed the new Tuskegee Airmen unit of black fighter pilots one year before his death. What he did not live to see, and what he had worked so hard to create, was a world where black airline pilots, and then black astronauts, were no longer unusual enough to attract special notice.

Indeed, this segue from theater to day-by-day function is a theme we also read in the women pilots of the late 1920s.[17] A pilot named Ruth Law had been present on that day when Harriet Quimby fell from her airplane. She went on to a spectacular career as a barnstormer, and, after she finally gave it up in 1922, she grumbled that: "Things are so proper now. . . . A pilot has so many rules . . . to follow. . . . I couldn't skim over the rooftops or land in the streets. . . . The good old days of flying are gone."[18]

I heard echoes of that same theme from my father, who last flew in 1919. He had loved the sense of flight that he felt in Sopwith Camels and Nieuport 17s. Then he had an opportunity to fly one of the fast and powerful new SPADs. He complained to me, "That wasn't any airplane; it was gun platform with a big engine. You didn't fly it, you just aimed it." The intimacy of the pilot with the surrounding air was already fading.

Quimby, Law, and Coleman were daredevils of the old school. The new generation of women fliers was still daring, but the old hedonism of flight was fading. These pilots began seeing flight in terms of function and purpose. Airplanes now had greater range, and flight was reaching into the far corners of the world. Barnstorming was first replaced with distance record setting, then with making flight into a functional means for transporting people and goods (more on that in chapter 11).

Women fliers were taking up these new roles in such remote regions of the planet as Africa and Australia. Jean Batten had been a three-year-old child in New Zealand the day Harriet Quimby fell from the sky. Batten learned to fly in 1928. Like Katherine Stinson 16 years before her, she had sold her piano to pay for flying lessons—like Stinson, she had honed her superior motor skills on the piano.

Stinson had gone from America to France to learn flying. Batten went from New Zealand to England. In 1933, she set a record flying from England back to Australia. Then she set a second record by being first to make the return trip. She crossed the South Atlantic Ocean in record time. She took hair-raising risks and set other records. Still, she eventually retired and lived to the age of 75.

The year Batten flew from England to Australia, a 17-year-old girl, Nancy Bird, was learning to fly in Australia. Within two years, she was the youngest woman ever to hold a commercial pilot's license. Bird did some racing, but she was less interested in record setting than in public service. A Catholic priest recruited her to help take medical service into the western Australia outback. Still a teenager, Bird became a lifeline to remote settlers—shuttling fresh fruit, bringing the sick to hospitals in Sydney, and earning barely enough to keep her airplane aloft by shuttling passengers.

In women like Nancy Bird, all the stunting, and the even the record setting, finally mutated into something far more serious. It may seem ironic to find flight maturing at the hands of a 19-year-old. But perhaps it takes the young to make sense of their forebears' still-half-formed ideas.

While the African American Bessie Coleman was doing stunt flying in America, the Anglo-African Beryl Markham was learning to fly in Africa.[19] Markham's maiden name had been Clutterbuck. She was born in England in 1902, and her father took her to Africa when she was four. She got the name Markham from her second husband.

Beryl Markham grew up on a Kenya farm, where she learned to hunt with African boys and was once mauled by a lion. Her schooling was minimal. She took up horse training in her late teens and flying in her late twenties. By then the beautiful Markham had married twice, mothered a son (whose father may have been the Duke of Gloucester), and was woven into the decadent, upper-class expatriate English life of pre-war Africa.

She was a friend of Isak Dinesen (played by Meryl Streep in the movie *Out of Africa*), but the friendship suffered when Markham took up with Dinesen's friend (played in the movie by Robert Redford). Markham flew airmail in Africa, rescued wounded miners and hunters in the bush, and spotted bull elephants for rich hunters. She was honing the use of the airplane as a functional tool but was nevertheless as serious an adventure junkie as Quimby or Coleman had been.

Her fame as a flier was established when, in 1936, she made the first east-to-west flight from England to the North American mainland: She wrote about her fears before that flight. "We fly but we have not 'conquered' the air. Nature presides in all her dignity, permitting us [to use her forces]. It is when we presume to intimacy, having been granted only tolerance, that the harsh stick falls across our impudent knuckles . . ."[20]

Her flight came late in the string of early transatlantic crossings that tested and probed Nature—each as dangerous as the last, and each still dependent on specially built airplanes. Markham's east-to-west crossing was the last ad hoc challenge to Nature that had to be made before commercial transatlantic traffic could begin three years later.

Nature rapped Markham's knuckles smartly. She ran out of fuel over open sea before she got to Cape Breton. Her Percival Vega Gull airplane sputtered and died just as land came into view. The plane upended in a peat bog, a few miles short of her intended landfall, and she stumbled out with a badly cut head.

But Markham's more remarkable feat was not in the air; it was in literature. Her one book, *West With the Night,* is a tour de force arriving out of some unexpected ether. She wrote the book about her life in Africa and her life in the air, soon after she had made the Atlantic crossing.

It is extraordinary writing by any measure. Hemingway said it made him ashamed of everything he had ever written. No doubt she had help from her third husband, writer Raoul Schumacher; but how much help? Rough drafts show editorial markings in both their hands. The problem is that Schumacher never wrote anything else to approach it, while Markham never wrote anything else *at all.*

So where did this masterpiece come from? I suspect that one huge factor was that Markham was a creative coiled spring, wound tight by life on the edge—a spring that uncoiled only once, leaving us all the richer for that one great whirl of expression.

If the lives of Markham and so many other early fliers were complex and shot through with ambiguities, even more was the life of the most famous woman flier of them all—Amelia Earhart.[21] Earhart was practically made for her role. She and Lindbergh were two peas from the same pod. Both were lean and shy, and you could have mistaken them for brother and sister.

In 1928, the year after Lindbergh flew the Atlantic, Philadelphia socialite Amy Guest asked publisher George Putnam to organize a transatlantic flight in her own Fokker Trimotor. Putnam found a pilot to fly it. Then he interviewed women to find, "[a] girl who would measure up to adequate standards of American womanhood." She would keep the flight log.

Earhart did her homework. Don't be too appealing; that might cause Putnam to be protective. Don't make too much of the fact you are also a pilot; they want only a second banana. She showed up looking just like Charles Lindbergh, kept her mouth shut, and got the job. She became the first woman to make the flight, and she stepped out of the plane into a media maelstrom that never ended for her.

Earhart was the hottest item in the papers, even if she had been only a passenger. Putnam became her publicist. He booked high-pressure speaking tours and fed stories to the press. Putnam made it hard on other women pilots. It was an odd symbiosis. He manufactured her fame, and she rode the wave until she began drowning in it.

For Earhart was driven by idealism. She had started out to be a poet. Then she learned to fly from a rare woman instructor named Neta Snook in California. By the time she rode across the Atlantic, she was a seasoned pilot, supporting herself as a social worker in Boston.

Then Earhart married Putnam in 1930, and it began as a marriage of convenience between two workaholics. She even made Putnam sign an agreement so she could dissolve the marriage after a year if she chose. Still, the marriage seems to have gained in meaning as the mad whirl went on.

For nine years Putnam managed Earhart, pushing an able pilot into the limelight over better fliers. Earhart finally flew the Atlantic solo in 1932. All the while she wished she had the mind-space to write poetry, but she never had time to meet her own standards. Instead, she used her bully pulpit to push things she believed in: women's rights, pacifism, and (above all) flight itself.[22] She was powerfully dedicated to all three causes.

Then she set out to fly around the world. If she had made it, the flight might have been forgotten. But she vanished at sea, and we spent the rest of the twentieth century wondering what became of this complex woman, who lived in a tangle of ambiguities. Was this last flight cover for naval spying on the Japanese? One theory says that Earhart used it

Amelia Earhart in a Cierva Autogiro. (Image courtesy of Stephen Pitcairn, Pitcairn Aircraft Company archives)

to vanish from the public eye so she could live out her life under another identity.

It was, undoubtedly, no more than bungled navigation; however, the dropping-out-of-sight theory reflects the sad truth of Earhart's plight. She was shy, bright, and caught in an ever-expanding web of manufactured success, and we are left to wonder what she might have said in her book of poetry.

In 1996, Jane Mendelsohn came out with a marvelous little book: *I Was Amelia Earhart.*[23] The first three paragraphs are as perplexing as they are beguiling:

> The sky is flesh.
>
> The great blue belly arches up above the water and bends down behind the line of the horizon. It's a sight that has exhausted its magnificence for me over the years, but now I seem to be seeing it for the first time.
>
> More and more now, I remember things. Images, my life, the sky. Sometimes I remember the life I used to live, and it feels impossibly far away. It's always there, a part of me, in the back of my mind, but it doesn't seem real. Whether life is more real than death, I don't know. What I know is that the life I've lived since I died feels more real to me than the one I lived before.

Perhaps I am so attracted to Mendelsohn's book because she voices my own question, "What did all that heroism mean?" Historian David McCullough offers a clue.[24] He identifies two odd attributes of early fliers. They looked good and they wrote. I go back to a photo of my father in a World War I flier's uniform. He was a handsome young man who went on to become a writer. I go back and look at photos of Harriet Quimby, Bessie Coleman, Beryl Markham, Amelia Earhart—all beautiful and literate.

An odd selection process is running here. Take Lindbergh: He wrote books, and his lovely wife, Anne Morrow Lindbergh (who also flew), created a body of literature quite apart from flight. The flier Antoine de Saint-Exupéry wrote many books, including that masterpiece of children's literature, *The Little Prince.*[25] He and the Lindberghs were close friends, drawn together by words as much as by flight. And listen to these words from Beryl Markham's *West with the Night.* She tells about leaving an African airport:

> We began at the first hour of morning. We began when the sky was clean and ready for the sun and you could see your breath and smell traces of the night. We began every morning at that same hour, using what we were pleased to call the Nairobi Aerodrome, climbing away from it with derisive clamour, while the burghers of the town twitched in their beds and dreamed perhaps of all unpleasant things that drone—of wings and stings, and corridors of bedlam.

Was prose like that borne out of literary genius or out of the experience of flight? My images of flight were powerfully formed by my father's story-telling. I lay in bed while he told me stories laced with images of turning and wheeling among pillars of clouds; of danger, fear, walking away from crashed airplanes; of the sounds and smells of flight.

And so they wrote. Nevil Shute wrote and Ernie Gann wrote. Amelia Earhart wrote prose when she wanted to write poetry. Saint-Exupéry wrote, "I saw the alchemy of perspective reduce my world, and all my other life, to grains in a cup." Lindbergh said that the airplane plunged him into the heart of the mystery of existence.

There can be no doubt that the airplane drove the prose. And the attractiveness? Maybe that can be explained by Abraham Lincoln's remark that, after a certain age, we bear the responsibility for our face. Maybe these people simply accepted that responsibility. I think Saint-Exupéry catches a wink of the meaning of it all when he describes a pilot on a night flight "falling into the deeply meditative mood of flight, mellow with inexplicable *hopes*."[26]

If the sky lured the disenfranchised in with its age-old promise of an escape into freedom, it rewarded those it drew with something more as well. If, at first, you think you are seeing a simple bargain being struck between exploiters and the exploited—men exploiting women for thrills, women exploiting opportunity to gain liberation—then look again. When you read the prose it generated, another theme entirely emerges. For early flight offered a transcendent, sensate experience. And that experience seems to have overpowered all else. That is what finally becomes clear only when we read what they wrote. Nothing as base as mere exploitation could have created all those haunting words, or that moral force, or such a sheer cry of pleasure.

Modern meant an explosive increase in radical new machinery, and new machinery is always tied to the metaphors we live by. The first and primary metaphor of flight is freedom, and it has been for millennia: "I'm free as a bird." "Once she learned math, she sprouted wings." "If I had the wings of an angel, over these prison walls I would fly."

But flight brought with it another metaphor that powerfully shaped *Modern*. That was the metaphor of speed and intensity. "My, he could run; he really flew!" "I can't stay, I have to take off." "'Was I speeding, Officer?' 'Hell, no, you were flyin' too low.'"

And there is no better place to begin that part of our story of flight than with the first great daredevil, Lincoln Beachey.

11

Flying Down to Rio

Imagine a scene on the shores of San Francisco Bay in 1915.[1] Fifty thousand people watch a diver from the battleship *Oregon* being lowered down into the water; then they wait. An hour later, the ship's winch begins turning. The crowd holds its breath until a primitive airplane finally emerges from the bay. Its wings are torn away, and still seated at the controls is the drowned body of Lincoln Beachey—until now, the greatest early daredevil of heavier-than-air flight.

The 28-year-old Beachey was a San Francisco native son—arrogant, uncouth, and made of pure brass. Eleven years earlier (and one scant year after the Wright brothers made their first powered flight) Beachey had made his own first flight. He was one of many teenagers who took a balloon ride across the Mississippi River at the Louisiana Purchase Exposition in St. Louis.

Flight was on everyone's mind at the world's fair that year, and it drew Beachey like flame draws a moth. At first, he himself piloted the small dirigibles that were becoming regular state fair entertainment in the Midwest. Then, in 1910, he enrolled in Glen Curtiss's flying school. He was a disaster as a pilot. He crashed, and he broke up equipment. Yet Curtiss recognized the fine madness of a real daredevil, and he hired Beachey to fly as a member of his exhibition team.

Beachey went on to perfect every kind of stunt. He flew down into the mists of Niagara Falls. He flew up to an altitude record of over two miles. He nose-dived from three thousand

Lincoln Beachey, from *The Am* *can Review of Reviews*, 1911

feet with his engine off while onlookers screamed, fainted, and vomited. He did one of the first loops in the air, and he did it hair-raisingly close to the ground. Prize money amply paid for any damage he did at the Curtiss Flying School.

Racecar driver Barney Oldfield had been the greatest daredevil of the times. Now Beachey laid claim to that laurel. It turns out that they shared the same agent. Like Oldfield, Beachey was a public-relations person's dream. He told reporters that he was consumed with remorse over deaths among pilots trying to outfly him. "Only one thing [drew audiences] to my exhibitions," he wrote. "It was the desire to see . . . my death."

In that age of flying box kites, nine out of ten exhibition fliers did, in fact, die. It was only a matter of time, and Beachey flew constantly. In 1913, his notices said, he had entertained 17 million people in 126 cities. Each routine was closer to the edge than the last. He finally laid off for a year, but then he returned to San Francisco with a new airplane to perform at the Panama-Pacific Exhibition.

Part of his routine was a series of loops—climbing and diving. This time he climbed and dove for the crowd faster than ever before. As he tried to pull out of one of those dives, the wings tore off his airplane with a sickening sound.

His funeral was a major event, with even the mayor of San Francisco serving as one of the pallbearers. School children made up a song for the occasion:

> Lincoln Beachey,
> Bust 'em green,
> Tryin' to go to Heaven,
> In a green machine.

Beachey had made his last headline, and, although others would replace him soon enough, his particular message lingered. He had understood, perhaps better than anyone before him, that the new flying machines were offering something beyond an escape.

If Shepherd, Quimby, Law, and others had shown us how flight serves as a metaphor for the door out of our various prisons, Lincoln Beachey had done a superb job of transmitting the airplane's other iconographic impulse—the pure, hedonistic urge toward speed and adventure. He tapped into that particular dimension of *Modern* in a way so raw and unvarnished—so explicit—that there could be no dodging it. We were entering a dangerous and intense new age. The new *aeroplanes* were exciting. They could kill us. And that is exactly why they were so much fun.

For two decades after Beachey was fished out of the bay, countless other flyers, cut from the same rug, continued in his footsteps. Commercial air service was seeking a solid footing, and those who followed Beachey would never coexist with it. We are back to the evolving iconography of flight, which, as it turns out, did not help this situation one bit.

A great confusion eventually grew up around the question "Is flight to be a door out of prison, a high road to adventure and death, or a new safe, rapid, and convenient form of public transportation?" In 1933, the RKO movie studios responded to that question with astonishing kitsch eloquence. They released the movie *Flying Down to Rio*—light as a feather, practically devoid of plot, and wonderfully rich in texture.

It included the first collaboration between Fred Astaire and Ginger Rogers, who were still only secondary characters. Their number *Carioca*, with its erotic and adventurous overtones, may have been the best thing in the movie; but it was only a sidebar to a huge musical celebration of flight. References to flying are incessant: a balloon landing in an outdoor nightclub, a marriage on a seaplane, a rejected lover exiting by parachute—and those scenes are from the straightforward parts of the movie.

The most startling number was the title song, "Flying Down to Rio." The premise is that the characters are prevented by local regulations from doing a hotel floorshow, so they decide to take their show into the sky. What follows is completely bizarre. Scantily clad dancers perform on the wings of flying airplanes in a scene so outrageous, so implausible, as to boggle the wildest imagination.

It was all done on a sound set, of course. No one could have carried it off in the sky. The movie simply picked up on daredevilry and made it into pure fantasy. No one was supposed to believe that dancers were really out there. This was about the way flight laid upon our imaginations in the years before commercial flight reached maturity.

Two years later, the Douglas Aircraft Company would produce its DC-3, and safe, reliable, commercial flight would become the norm. But, for the moment, those dancers linked air travel to the hair-raising county-fair stunt of wing-walking, something we had all watched. I have friends who saw people die trying to do it.

The message of that movie scene had a wonderful, illogical logic. It told us that danger was fun, and that danger was really perfectly safe after all. If passenger air travel was about to become safe and regular, wing-walking and daredevilry were not quite ready to leave the fairgrounds and become the stuff of myth—the stuff of my father's story-telling.

My father had, by the way, cursed himself all through my childhood for failing to appropriate one of the surplus fighter planes that were being piled up and burned after the German surrender. He wished he had had the nerve to fly off to the Eastern front to help fight the Bolsheviks. He had arrived in France in 1918 and failed to reach the front in time for combat. A huge dimension of danger and excitement had been withheld from him, and his personal sense of loss was immense.

He had, I suppose, failed to live up to that word *daredevil,* which so lay upon our minds in the early twentieth century. Coined in the late eighteenth century and given increased use during the nineteenth, *daredevil* now became an integral part of the vocabulary of *Modern.* By the time Beachey made his not-unexpected exit, racecars had been moving at far higher speeds than airplanes from the beginning.

The great around-the-world automobile race (chapter 8) took place in 1907, just before the first heavier-than-air aerial showmen took to the

Racecar illustration, from a painting by William H. Foster
in the June 1913 *Century Magazine,* p. 222.

sky. The first around-the-world flight would have to wait only another 17 years (more on that in a moment). Automobile drivers were the heroes of the day in 1907. They owned the term *daredevil*, but they would soon yield it to the new suicidal species of aviators.

It was clear even then that the airplane was destined to become a routine mode of travel. Yet one look at the airplanes available at the end of World War I told us how unsuited they, and the pilots who flew them, were for bringing such travel about.

One of the more promising American leftovers from World War I was the deHavilland DH-4. It was an odd duck among airplanes of that period. Its story, told by a National Air and Space Museum director, Walter Boyne, begins in 1917. Boyne asks an unexpected question in the introduction to his book about the DH-4: "Why was the United States, which had given birth to the airplane ... so absolutely indifferent to its development [before World War I]?"[3]

By "United States" he means the U.S. government, for the American public had been far from indifferent. However, by the time the war began, Germany had already spent $45 million on military aviation, Russia $23 million, and France $13 million. The American government had spent a scant $400,000 on military aircraft.

America was casting its lot with railways and highways, while pilots all over were busy shaping the airplane's role in defining *Modern*. America's primary prewar military airplane, the Curtiss Jenny, had actually gone to war as early as 1915—not against Germany, but against the bandit Pancho Villa in Mexico. Eight Jennies and ten pilots were no match for the harsh Southwest. They were not meant to stray more than about 50 miles from their bases. They got lost, they crashed, and General Pershing never did catch Pancho Villa.

When America entered the war in 1917, it appropriated $13 million for airplane development, and went to the European airplane builders looking for designs. The British deHavilland company had built a series of airplanes, the best of which was the Rolls-Royce-powered DH-4. The British sent us one to use as a template.

In the end, we spent $640 million on creating our own version of the DH-4. We switched from the British system of handcrafting airplanes to our automobile production methods. We developed our own so-called Liberty Engine—a large, four-hundred-horsepower, water-cooled, V-12 machine. We renamed the finished airplane the "Liberty Plane." The first of our DH-4s had so many design bugs that fliers called them "Flaming Coffins." Those bugs, however, were soon set right.

My father joined the Army Air Service in 1917, and he trained in Jennies. The DH-4 reached France in May 1918, about the same time he did. He never piloted one, but a friend did give him a ride over the trenches in one after the armistice (see chapter 14). He used to tell me about its early reputation as a funeral pyre. Still, the DH-4 had, by then, become a very solid airplane. Its real flaw was that it was woefully unsuited to combat. It was far too ponderous to dogfight the German Fokker triplanes and bi-planes, and it carried too small a load to be a useful bomber.

The DH-4 was a clear matter of too much, too late. However, thousands of DH-4s (and an even larger number of Liberty Engines) were left over in America on Armistice Day. The big biplane had a tubular steel frame as well as a fine engine, and, when it crash-landed, its pilot walked away alive more often than not.

Right after the war, the army began its airmail service with DH-4s. DH-4s with skis served in the Arctic. Engineers modified them in every way imaginable. The Liberty Engine was so good that people relied on it instead of moving forward. In the end, it probably inhibited American aircraft engine development.

The DH-4 occupies a strange place in aviation history. Although it had been no match for the famous World War I combat airplanes, it was still in service when all that remained of Fokkers and SPADs were lingering leg-ends of speed, danger, and exhilaration. It formed one of the bridges be-tween Lincoln Beachey and commercial air service. In doing so, it dramatized the mixed message we were receiving after World War I.

A DH-4 as shown in the April 1919 *Century Magazine*, p. 844.

Flying Down to Rio was still delivering that message in 1933. The air-planes upon which the chorines danced were all single-engine machines. Indeed, one was remarkably similar to the DH-4. The airplanes in the director's mind were still those that had been used by barnstormers. It would have undercut the excitement of the movie to use the new trans-port airplanes that were coming into use for staid airline service.

The long-range passenger-carrying potential of airplanes had been made clear much earlier. Taking their cue from automobiles, airplanes had made their first round-the-world trip nine years before *Flying Down to Rio*. In 1924, three years before Lindbergh's transatlantic flight, the

Army Air Service took on the project of circumnavigating the earth in airplanes, and the systematic way they went at it makes me think of NASA's lunar program.[4]

The Douglas Aircraft Company was asked to make four special biplanes: the *Boston*, the *New Orleans*, the *Chicago*, and the *Seattle*. Each had a 450-horsepower Liberty engine, and these big airplanes could carry enough fuel for a 2,200-mile flight. They could be fitted with either pontoons or wheels.

The army used the flight from Santa Monica to Seattle as a shakedown trip while it provisioned supply depots for the four planes in locations around the world. The official flight began on April 6, 1924. The airplanes headed up the Canadian coast toward Alaska, stopping at Prince Rupert, Sitka, Chignik, and Dutch Harbor in the Aleutian Islands.

The *Seattle* suffered minor damage in its first landing. Then its engine had to be replaced. Finally it crashed in west Alaska. The pilot and copilot survived the first night by burning the wings as fuel and huddling in the baggage compartment. They eventually managed to walk a little further westward to the Bering Sea, where a cargo ship rescued them.

From Attu, the remaining fliers meant to go to Kamchatka Peninsula, but the newly formed Soviet Union wouldn't permit it (Russia itself was struggling to set long-distance records). So the three airplanes continued to Japan's Kurile Islands. They reached Tokyo on the 49th day, and they had, technically, been the first people to fly the Pacific.

From there they went on to Saigon, Bangkok, Rangoon. At Calcutta they switched from pontoons back to wheels. They replaced all three engines in Karachi. From there they flew to Baghdad, Aleppo, Constantinople, Bucharest, Vienna, Paris, and, finally, London. Next they had to cross the Atlantic—*against* the westerlies and three years before Lindbergh. To do so they used an island-hopping strategy.

After a stop in the Orkney Islands, the *Boston* crash-landed near the Faröe Islands and sank during the rescue attempt. The *New Orleans* and the *Chicago* continued to Iceland, then Greenland, where they again replaced their engines. They reached the mainland at Icy Tickle, Labrador, and flew down eastern Canada to Boston.

The trip ended with a triumphal tour of America. The two remaining airplanes reached Seattle on September 28, 1924, after a 175-day journey. Their time in the sky had totaled 15 days, which meant an average in-air speed of 70 miles an hour. Their overall speed for the 27,000-mile odyssey, however, had been a scant six miles an hour. They could have done it much faster in a sailing ship.

All eight pilots survived, and you can see the two surviving airplanes today in Dayton's Air Force Museum and in the National Air and Space Museum. Like that auto race 17 years earlier, the feat convinced us all that this new machine would eventually become an instrument of wide-spread public transportation. Like the auto race it had been the work of people who were both daredevils and pioneers.

As the trip bore fruit in the mid-1930s, a remarkable book appeared that said, in effect, that we needed to look to the airplane's functional evolution if we were to have any hope of understanding *Modern*. In 1935, the famous architect Le Corbusier offered us a picture book entitled *Aircraft*.[5] Always the iconoclast, Le Corbusier added a subtitle when he was finished. It was *L'avion Accuse*. He explains himself in the foreword:

> The airplane is an indictment.
> It indicts the city.
> It indicts those who control the city.
> By means of the airplane, we now have proof, recorded on the photographic plate, of the rightness of our desire to alter methods of architecture and town-planning.

Le Corbusier's real name was Charles Edouard Jeanneret. He was born in Switzerland in 1887, and he trained in art, engraving, and architecture.[6] He emerged as an artistic revolutionary in Paris just after World War I. He started an avant-garde magazine, *L'Esprit Nouveau* (The New Spirit), and in it he took on architecture, science, technology—even music. Darius Milhaud was one of his collaborators.

Le Corbusier waged war on every part of the establishment. Education and cities were his favorite targets. He insisted that "men have built cities for men, not in order to give them pleasure, to content them, to make them happy, *but to make money!*" He told us that the old orders of architecture were hopelessly inhumane. They had forgotten function and human need.

As an architect, Le Corbusier began building in rough-hewn concrete. It was lovely stuff—light and open. But as he built buildings, he also built a complex, often murky, new dialectic of architectural values. The airplane finally became the metaphor by which he could explain, in terms we can understand, his belief that the world had been built ill. The airplane was a wholly new form. It was pure function. It derived its beauty from function. The airplane, he said, "embodies the purest expression of the human scale and a miraculous exploitation of material."

The pictures he shows us are beautiful. Such wild machines filled the skies before World War II: graceful gliders, lumbering transports, exotic racing planes, amphibians, airplanes with three engines, or with eight.

"No door is closed," he says. "Everything is relative. If a new factor makes its appearance, the relation alters. It is a question of harmony. 'In aviation everything is scrapped in a year.'" And to get from here to there, an airplane simply flies in a straight line.

For Le Corbusier, machinery and craftsmanship are the one truth in a world full of lies. Machines are truly humane, but we don't know machines. He says, "Countries lack the **harmonizing influence** which would make the **humane** beauty of modern times palpable."

Those ideas take on sinister overtones in hindsight. Le Corbusier went into the sky, looked down on the city, saw dysfunction, and went back to his drawing board to rectify it. What he produced were city plans that might have been drawn up by King Camp Gillette (see chapter 7). He wanted to create a degree of organization that veered dangerously in the direction of the fascist thinking that was laying its hold upon Europe.

But look at the photos in his book and you see a terrible beauty in those airplanes. Each is different; all are transient. As each struggles in its own way with the rigors of carrying us into an unwelcoming sky, we catch a glint of his meaning.

The airplanes of 1935, buoyant and fluid, did indeed indict the static cities below them. Le Corbusier showed us what he meant by "harmonizing their beauty and making it palpable to us." When he concluded that the "pre-machine civilization is finished," he captured the impulse of *Modern* with terrifying accuracy. We look around us today and see that he was correct.

Architecture actually began responding to the airplane before Le Corbusier. Shortly after World War I emerged something called the *airplane bungalow*. This form of house had a wide, flat roof that resembled an airplane wing, and it was topped by a cockpit-like partial second story. It combined vestiges of flight with the clean linearity of art deco.

One finds those same themes in Le Corbusier's architecture, but rendered in a surreal way. His reinforced-concrete Chapel of Notre Dame du Haut in Ronchamp, France, was finished in 1956 (in the dog days of *Modern*), with a roof that still calls to mind the wing of an old airplane.

An airplane bungalow, ca. 1922, in Los Angeles, California.
(Photo by Margaret Culbertson)

Chapel of Notre Dame du Haut in Ronchamp, France, designed by
Le Corbusier, 1956. (Architecture student photo, University of Houston)

Let us return, then, to the movies, for they show us how the airplane was
strengthening its hold on our psyche. The same year that RKO released *Flying Down to Rio*, Warner Brothers issued the movie *Central Airport.* Two real
pilots directed it, and it told the story of two fictional pilots—brothers in
love with the same woman. The action moved from one embryonic airport
to the next, all the way from Los Angeles to Havana.

Where *Flying Down to Rio* used fantasy to represent the new airplanes,
Central Airport used realism. Which movie provided the greater insight
is a toss-up. *Flying Down to Rio* laid our ambivalences out on the table,
while *Central Airport* provided an accurate look at the texture of early
commercial flight in 1933.

Four years earlier, the Curtiss-Wright Corporation had published a
description of the new Curtiss-Wright system of airports.[7] It explains
what we see in the movie *Central Airport*, which seems (for all its accuracy) to have been misnamed. There was nothing very central about these
first airports. The book gave architectural renderings of 12 completed
airports, and it suggested a vast, decentralized array of airports and infrastructure yet to come. Runways are one-half to two-thirds of a mile
long. Terminal buildings are one story high.

Then, in 1933, United Airlines continued the theme in its guidebook
Airways of America: United Airlines.[8] The United Airlines book describes
the route from New York to San Francisco. It details all the physical geography of America along the way. There were still no pressurized cabins, and one flew beneath most of the cloud cover. In those days one *saw*
the earth below.

Ford Trimotor, Evergreen Flight Museum, McMinnville, Oregon.
(Photo by John Lienhard)

New York to San Francisco was a two-day trip with endless stops along the way. The guidebook was meant to keep a passenger occupied with mountains and prairies; nimbus and cumulus clouds; glacial, alluvial, and volcanic deposits—America opening up below.

Both books show America caught up in very rapid change, yet neither correctly tells us where the change is headed. The whole system is being shaped to the one airplane that has become the workhorse of commercial travel. It is the 1926 Ford Trimotor, and it, too, is an American improvement over a European design, the Fokker Trimotor. It would be America's basic passenger carrier until Douglas made the DC-3 in 1934.

The Trimotor carried 11 passengers unless there was a stewardess; then it transported only ten. It had one 200-horsepower engine under each wing and one in the nose, a flight ceiling of 16,000 feet, a cruising speed of 107 miles an hour, and a range of 570 miles.[9] (The relatively short range of the Trimotor was the reason that the Curtiss-Wright Corporation envisioned such a closely spaced network of airports across America.)

United Airlines and the Curtiss-Wright Corporation took great pains to tell the public how safe and pleasant flight had become. The movie *Central Airport,* on the other hand, was blunt about the danger of flight. It showed a total of three airplane crashes. In fact, the development of the DC-3 was spurred by the death of football hero Knute Rockne in a Trimotor crash. The sky was still a dangerous place. It was still romantic. The hard earth was still down there, waiting for us to fall from the sky.

Today we enter a sealed environment in one city and leave it in another. At just under the speed of sound, 35,000 feet up, I can read a novel. Seventy years ago we still felt kinship, not just with Lindbergh and Earhart, but with Lewis and Clark as well. To fly across America was to discover it. We would watch the sparse population below putting in orchards and wheat fields while they interspersed them with airports and two-lane highways.

The same year that the Curtiss-Wright Corporation published its book on domestic airlines, it also proposed a radical solution to the essential problem of transoceanic flight—the *diminishing-returns* problem of fueling long-distance flight.

That problem loomed very large then, and it is still present today. When Lindbergh flew the Atlantic in 1927, he did so by turning a specially designed Ryan monoplane into a flying gas tank. Over half its take-off weight was gasoline. When Burt Rutan's airplane *Voyager* circled the world nonstop in 1986, its take-off weight was *80 percent* fuel. Fuel in a three-stage rocket system greatly outweighs the space shuttle that it carries into orbit.

Any long trip that must be made without refueling leads to a rapidly diminishing return of payload for the fuel required. That issue underlay the Curtiss-Wright proposal that America be served by a system of small airports, each within a few hundred miles of the next. Longer flights, while *possible*, were not *feasible*, and regular transoceanic air service appeared to be out of the question.

Author Steward Nelson tells how, in 1929, the Curtiss-Wright Corporation also picked up on an idea being promulgated by Canadian/American engineer Edward Armstrong.[10] Armstrong had begun selling his solution to the problem as early as 1913. He hoped to build strings of floating airports, called *seadromes*, across the Atlantic.

A seadrome was to weigh 50,000 tons and have an 1,100-foot-long deck. Its flotation system would extend about 180 feet into the water. To hold it in place, Armstrong went to John A. Roebling and Sons, the company that had built the Brooklyn Bridge 40 years earlier (see chapter 2). They designed a deep-water anchoring system for Armstrong.

Each seadrome would have a 40-room hotel, a café, a lounge, and other amenities. On October 22, 1929, the *New York Times* announced that construction of the first seadrome would begin within 60 days. Seven days later was Black Tuesday, the day the stock market crashed and the Great Depression began.

Armstrong struggled on. He almost gained PWA support for the project. In 1934, he briefly had Roosevelt's interest. All the while, airplane engines were improving, and the feasibility of nonstop transatlantic service increased. In fact, the great Zeppelins already provided regular transatlantic service. Still, Armstrong's dream lasted all the way through World War II.

Driven by a new kind of need, the Japanese built a one-kilometer-long floating airport in 1999 and called it *Megafloat*. They meant to create offshore airports to avoid consuming the precious land of crowded coastal cities. That was the same thing that big seaplanes had achieved until they were doomed by their inherently high-drag, low-speed profiles.

Armstrong's seadromes are still with us, but the concept has now split in two. In addition to considering the creation of floating coastal airports, we have turned the seadrome into a commonplace reality in the big floating offshore oilrigs. By the year 2000, over 25 percent of America's domestic oil was being provided by such platforms.[11]

Our search for the end of daredevilry and the beginning of settled, safety conscious civil air travel thus blurs. *Flying Down to Rio* was already an anachronism by 1933, even though it embodied a powerful surviving impulse. The venerable magazine *Aviation Week,* begun in 1916 as *Aviation and Aeronautical Engineering,* offers another window into the transition.[12]

In a 1919 article by Orville Wright, "The Future of Civil Flying," one can read frank puzzlement over the direction of aviation. Like his brother, who had said the primary purpose of flight would be sport, Orville felt that civil flying would never be so much about moving passengers as it would be about play.

He acknowledged that one could get from Dayton to New York four times faster by plane than by train, but he argued that it was not safe to do so because one passes over mountains and cannot make emergency landings. Maybe the government would eventually build landing strips in the mountains, Wright mused, but intercity flight would surely remain a private matter more than a commercial one.

The most important need, according to Wright, is that airplanes be designed to have slow landing speeds. After all, the length of a landing strip rises as the square of the landing speed. If we could cut landing speeds to 30 miles an hour, we could fly far more safely. Then flight would take its place as a widespread sport.

Two years later, in 1921, the magazine shows a photo of the giant new Italian seaplane, the Caprioni-60. Its six wings and eight engines were to carry one hundred passengers at a lumbering 80 miles per hour. Its range was to be four hundred miles. With its forest of struts and wings, it looks more like an unfinished balloon-frame house than an airplane. Later that year, on its maiden flight, the C-60 got 60 feet into the air before it crashed into Lake Maggiore.

By 1931 rudimentary commercial flight was finding its footing, and Amelia Earhart wrote an article as vice president of Luddington Airlines. Luddington had already flown over a half-million passenger miles between New York and Washington. Only 12 years had passed since the publication of Orville Wright's article, but her message is much different.

Airlines must figure out how to keep their schedules. Another of their big problems is learning how to manage ticket sales. (This was the world before computers.) And, of course, airlines need to keep their prices low.

Earhart brushed off the pernicious problem of airsickness, saying it afflicts only 5 percent of travelers. More women suffer than men, but they're more stoic about it.

The essential difference between Wright and Earhart is that he was still reading a crystal ball and reflecting his dreams as an inventor. Although Orville Wright outlived Amelia Earhart, she was a *user* of the airplane, not an inventor. Her feet were firmly planted in the epoch of *Modern,* and she dreamt about where the invention might take us. Since she let the machine work through her to define itself, she saw it more clearly.

By 1942, America was at war again. Based upon the accumulation of civilian experience, the nation's large bombers could now navigate the oceans regularly. Was the daredevil impulse being resuscitated by war? Not really. Heroes in airplanes were being killed, but not for the fun of it.

I first heard the radio serial *Hop Harrigan* that year. Each episode began with a voice: "This is CX4 to control tower, CX4 to control tower. Hop Harrigan, America's ace of the airways, coming in!" VaarrooOOOM! Harrigan was an old-school daredevil who came to radio out of the comic books. He was not an Air Force pilot. He still risked his neck for excitement.

The *Hop Harrigan* comic strip had begun in 1939, and the name was a clear homage to real-life hero Wrong-Way Corrigan. Douglas Corrigan had made himself into an American folk hero, just ten months before the comic strip appeared, by flying the Atlantic, supposedly by mistake.[13]

Corrigan was the last of the early glory-seeking fliers, and the year 1938 found him flying a single-engine 1929 Curtiss-Robin that he had bought secondhand for $325. No mean mechanic, he had rebuilt the airplane and modified it for long-distance flight. Like Corrigan himself, the airplane was a holdover from earlier days. Lindbergh had crossed the ocean 11 years before, and many others had followed suit. But it was still not a trick you would try in any regular, factory-ready airplane.

Corrigan flew nonstop from California to New York and then filed plans for a transatlantic flight. Aviation authorities denied the flight plan; the seeming piece of junk that he was flying had no business challenging the Atlantic! So he gave up and filed a plan for a nonstop return trip to California.

He took off at dawn on July 17, 1938. A few puzzled onlookers watched him do a 180-degree turn and vanish into the clouds. Twenty-eight hours later he stepped out of his plane in Dublin, Ireland, saying, "Just got in from New York; where am I?"

The authorities suspended Corrigan's license while he stuck to his story: His main compass went out and he'd misread his backup compass in the dark. He didn't realize he was mixing the point and the tail of the compass arrow until the sun rose the following day. He really had meant to fly to California.

The only people who were fooled were people who wanted to be fooled. Corrigan was the lingering aftertaste of our love affair with daredevilry and the perfect anti-hero—the little guy in his homemade airplane who had beaten the system.

By the time the ship carrying Corrigan and his crated plane got back to New York, the suspension had been lifted. The city gave him a bigger parade than it had given Lindbergh, and he starred in an RKO movie about the exploit, *The Flying Irishman.*

The next year saw the first commercial transatlantic airline flight. The year after that saw four-engine bombers being ferried off to war in Europe. Corrigan became a test pilot while I listened to *Hop Harrigan.* Of course Harrigan's plot material gradually veered off toward the purposeful business of fighting World War II. It could no longer be ignored.

When Corrigan was 81, his old Curtiss-Robin came out of mothballs and went on display at an airshow. Corrigan, who had grown reclusive, was suddenly so enthusiastic that the promoters hired a guard for the plane, lest he try to take off one more time. That was in 1987. When he died in 1995, he had engraved the expression *Wrong-way Corrigan* upon the English language.

Charles Lindbergh, Corrigan's boyhood hero, lived only to the age of 72, but that was long enough to bring him far beyond these impulses of *Modern.* Historian Leonard Reich tells how, much more than Corrigan, he left his own *Modern* behind.[14]

Not long before he died in 1974, Lindbergh wrote the foreword for a book about the Tasaday tribe in the Philippines—a tribe whose way of life seemed, at the time, to be under assault. Lindbergh wrote: "During decades of [flying we crowded ourselves into] cities that spread like scabs across the landscape. Obviously an exponential breakdown in our environment was taking place; and, just as obviously, my profession of aviation was a major factor in that breakdown. Aviation had opened every spot on earth to exploitation."[15] He had come very far from his youth.

Raised in a small Minnesota town, Lindbergh had a father who had been a progressive congressman and a mother who was a chemist who drew him into technology. When he was 11, she took him to Panama to see the new canal being built there. Lindbergh tried to take up engineering, but he didn't have the patience for it. He dropped out and then enrolled in a flying school. For a while he barnstormed, and, before he was 20, he had set his heart on winning the $25,000 Orteig Prize for flying nonstop and solo from New York to Paris.

In 1927, of course, he won the prize and the fame that went with it. By the mid-1930s he was consulting in Nazi Germany. He was deeply impressed with Hitler's technocracy, even if he distrusted the politics. Back

Charles Lindbergh and the *Spirit of St. Louis*.
(Stereopticon image courtesy of Margaret Culbertson)

in America, he preached defensive isolationism. When World War II began, he offered his services to the army, which, by then, saw him as pro-Nazi and refused to take him. Lindbergh went instead to industry. First he helped Ford make B-24 bombers. Later, he was almost killed test-flying a P-47 for Republic Aircraft.[16]

By the age of 42, Lindbergh was a "civilian" test pilot in the South Pacific. He combat-tested P-38s and Corsairs. He flew 50 bombing and strafing missions. He even shot down a Japanese plane. His work changed the tactics of low-altitude combat. Just as Hop Harrigan had had to mutate from daredevil to functioning member of the war effort, Lindbergh mutated as well.

In his 36th combat mission, as he flew over Palau in the South Pacific, a Japanese Zero got him in its sights. He wrote, "I can see the cylinders of the Zero's engines, feel its machine guns rising into line behind me. It is too close to miss. I pull my elbows inside the armor plate and brace for the impact of the shells."[17] He was saved moments later by another airplane. But this was no longer the voice of a barnstormer; it was the voice of an authentic hero for whom life had become precious.

After the war, Lindbergh influenced the development of civilian jets, but he also wrote the book of essays from which I just quoted. In it he expressed the conflict he saw rising between the benefits and the dangers of modern technology. His defining moment came in 1955 when his wife, Anne Morrow Lindbergh, wrote *Gift from the Sea*.[18] She described a simple life on the seashore and contrasted it with, in her words, "the curtain of mechanization [coming] down between the mind and the hand." He read that as an accurate criticism of the life he had spent pushing the machine ahead of its human users.

After that, Lindbergh became less visible. He worked with quiet intensity trying to implement sane environmental policies. He pushed for the creation of the Congressional Office of Technological Assessment. In fighting for the survival of that Tasaday tribe, he may have been caught in a great hoax by which the Philippine government was trying to grab land for itself.

However, at the end of a long and complex life, Charles Lindbergh was still doing what too many of us stop doing in our later years. Right up to the end, he was still struggling to get it right. In his last book, published posthumously, he says something that clearly places him on the other side of *Modern*: The ultimate exemplar of *Modern*, Le Corbusier, had said, "The bird's eye view . . . now sees in substance what the mind formerly could only subjectively conceive."[19]

But Lindbergh, very soon before his death, recalled visiting the ninth-century Chapel of St. Gildas on an island off the coast of Brittany, in 1938. First, like Le Corbusier, he says: "A decade before, when I was flying from America to Europe in my *Spirit of St. Louis*, I had looked down on the Atlantic and wondered what shapes and contours were masked by the sameness of its surface. . . . the sea maintained its dignified aloofness."

Now, however, he is on the ground, walking the beach, and he has something much different to tell us: "At the edge of Saint-Gildas, each fastly ebbing tide opened the ocean's threshold, let you step into a strange and foreign realm. Fish, camouflaged by weeds, hung motionless in crystal pools. Green, protoplasmic masses lay inertly on the stones. A tentacle from a small squid flashed out.[20] The game had changed. We had left barnstorming, just as we had finally left tailfins. The second time the world had erupted in total war, we emerged looking at the world through different eyes. Lindbergh was modern no more.

The young Lindbergh had been formed by powerful forces, which, I believe, pulled him (pulled us all) from one axis onto another, when he was young. Perhaps Lindbergh's generation, more than others, was ready for realignment in the 1950s.

In any case, we need to look more closely at the forces that affected the young boys of *Modern* so strongly. What was it that told Lindbergh in such irresistible terms that he should become a daredevil and risk-taker? And why, at the end, did he swing so far back toward the life of a philosopher, mystic, and environmentalist? We look next at a boy's life in the early twentieth century.

12

A Boy's Life in
the New Century

L et us begin with a scene in the late 1930s. I am seated at a gas-fired
iron stove in the basement laundry room of my house. I hold a
small iron ladle over the flame. It is half-filled with lead scraps that
I scavenged under the linotype machines at the St. Paul Dispatch and
Pioneer Press newspaper building. I found the lead when my father, writer
and sometime science editor at the paper, took me to see the machines.

The gas flame is turned up high, and the lead has become a silver
pool, partly covered with an oxide scum. In my left hand, I hold the
musket-ball mold that belonged to my great-grandfather in his days at
Sutter's Fort (described in chapter 1). It's shaped like a pair of pliers.
Each side of its mouth is half of the mold. I keep the halves together by
gripping the handles tightly.

Then I pour lead into the sprue hole of the mold, filling it to the top.
The lead solidifies right away, and I open the mold. The exposed casting
sticks to one side of the mold. I tap it on the concrete floor, and the finished musket ball tumbles out with the tapered sprue piece still attached.

Musket-ball mold showing
a freshly cast ball and
sprue piece.
(Photo by John Lienhard)

I have many uses for this homemade artifact. If I remove the sprue with a hacksaw and sand the spot smooth, I have a particularly vicious form of slingshot ammunition. The slingshot is also homemade. I build it from the Y-shaped branch of a sapling. For the powerful driving engine of this weapon, I find a discarded automobile inner tube and slice it (as one might slice a loaf of bread) into large, half-inch-wide rubber bands. The resulting device can do the same damage as a musket with a light power charge. I've made a mean piece of artillery.

The other use for that musket ball has equal potential for harm. This time, I leave the sprue piece in place and use it to attach a three-foot length of strong kite twine. Out in the street, I whirl the ball at the end of the string, faster and faster, and release it at just the right point. It flies vertically into the sky and I watch until it vanishes from sight. Then I back under a tree and wait for its return to Earth. There is, at length, a dull thud in the street, near where I stood when I released it. I go to investigate, pick it up by the string, and admire my now-flat-on-one-side musket ball.

Next, instead of trying for height, I try for distance. This time, I release it at a 45-degree angle, hoping that my luck holds and I do not hit a window or a person over in the next block.

One day I found a substantial length of discarded lead water pipe and went into full production of musket balls. By the time I had made seven or eight of them, the mold became too hot to hold. I took it to the sink and ran cold water over it, then went directly back to casting without bothering to dry it out. (I figured the molten lead would dry it for me.)

When I poured the lead in this time, the mold exploded. Molten lead blew outward in all directions, spattering my face and narrowly missing my eyes. To this day, lead is embedded in the ceiling of that basement room. I was hurt and frightened, of course, but ongoing injury was part of the game.

The most lasting effect of the accident was *mystification*. What on Earth had happened? My bafflement kept nettling me. I developed a gnawing interest in heat and heat phenomena. I became an engineer, studied thermodynamics, went on to graduate school, did my M.S. and Ph.D. theses on thermodynamics. I became a thermodynamics teacher and wrote books on both thermodynamics and heat transfer.

I also became deeply involved in research on superheated water.[1] I found that water can be heated more than 400°F (222°C) beyond its room-pressure boiling point before it becomes irredeemably unstable and has to explode. I found that the destructive power of exploding water rises as the square of the number of degrees to which it is heated beyond its boiling point.[2]

At last, I knew exactly *how* that mere drop of water in my old musket-ball mold had had the potential for nearly blinding me. By this point I

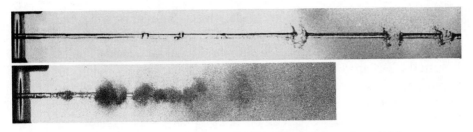

Two water jets issue from the 3/32-inch orifice on the left, at about 110 feet per second. The upper and lower jets are 80°F and 90°F above their boiling point. Both begin exploding after they leave a high-pressure environment.

had also learned that the real threat to a nuclear power plant is not a *nuclear* explosion, but a *superheated water* explosion. I worked for a decade on the effective means that have been created for protecting such plants against the danger of such thermo-hydraulic explosions.[3]

And yet my purpose is not to write a concise story about how one formative event set my life course. For all its importance to me, it would be silly to give that much weight to an explosion in my laundry-room laboratory. A hundred other boyhood adventures, and near misses, were also shaping my life.

Boys everywhere across America were up to similar mischief. We all felt the thrill of knowledge passing through our fingers and into our brains. Curiosity grew with each failure, and confidence grew with each success. Now and then, nature rapped our knuckles—kept us in check—but only barely. I, for one, was very lucky to come away with nothing worse than scattered scar tissue.

The burgeoning world of *Modern* had systematically tutored us all in these rites of boyhood. Naturally, my parents told me to be careful. They were no less concerned with my well-being than any parent might be today. However, they were creatures of *Modern*, and they let me go with the flow of the early twentieth century.

We need to look for the subterranean drum upon which this beat was carried. One very important source was a special literature being created specifically for the boys of *Modern*—from the late nineteenth century into the 1950s.

As an example, I offer you *The Boy Mechanic: 700 Things for Boys to Do.*[4] This book was put out by *Popular Mechanics* magazine in 1913, a decade after the Wrights' first airplane was built and a year before World War I. Ford had just begun mass-producing Model Ts. Telephones were now in general use (although they were still powered with large dry-cell batteries from the hardware store), and it would be over 20 years before the electric refrigerator reached middle-class homes.

The frontispiece of *The Boy Mechanic* shows a boy ready to step off a cliff in a glider that he has built according to the instructions on page 171.

The glider is rather like the ones that flight-pioneer Otto Lilienthal had flown until 1896, when he fell to his death in one. There were few lawsuits in 1913, and OSHA did not exist. If you read this wonderful old book and then killed yourself, it was your carelessness, not the publisher's. You knew you had better make the joints secure when you built the wing. I shall quote the full extent of the flight training provided by the book:

> You will find that the machine has a surprising amount of lift, and if the weight of the body is in the right place you will go shooting down the hillside in free flight. The landing is made by pushing the weight of the body backwards. This will cause the glider to tip up in front, slacken speed and settle. The operator can then land safely and gently on his feet. Of course, the beginner should learn by taking short jumps, gradually increasing the distance as he gains skill and experience in balancing and landing.

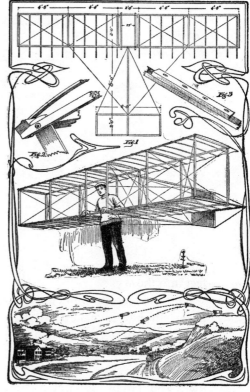

How to Make a Glider
(See page 171)

Frontispiece of *The Boy Mechanic*. Take particular note of the art nouveau décor woven around the images.

The writer finishes with a laconic warning: "Great care should be exercised in making landings, otherwise the operator might suffer a sprained ankle or perhaps a broken limb."

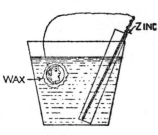

One section shows how to counterfeit a penny: First make a wax impression of the penny, then coat the wax with black lead. Finally, electroplate the penny into the wax mold. This is done by making a battery cell from a strip of zinc in an electrolytic solution. The text does not speak explicitly of counterfeiting pennies, but rather of replicating small objects in copper. The implication, however, is unmistakable.

The book is full of electrochemistry. It also tells how to generate hydrogen or acetylene and how to develop and print one's own photographs. It is strongly inclined toward things that explode. My favorite among these is the Fourth-of-July Catapult. It turns out to be nothing less

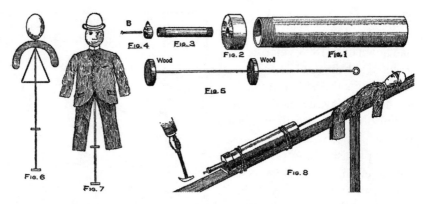

This was entitled "Homemade Cannon Which will Hurl a Life-Size Dummy 100 ft. through the Air."

than a pipe bomb capable of flinging a life-sized mannequin a hundred feet into the sky.

The mischief continues for 460 pages: how to play magician and levitate a lady, how to make a large ball appear to roll uphill, how to make a crossbow or to start a fire with a lens shaped from ice—in short, how to live on the edge.

The more sedate arts and crafts are certainly represented—how to make a lamp, a tie rack, or a coin purse. But they were not the reason for the popularity of such books with young boys like me. We read them because they told us how to engage all that delicious danger. This book had 800 pictures, and I find only one girl engaged in a craft. She's making a decorative lampshade, not a pipe bomb.

While there was nothing exceptional about *The Boy Mechanic,* it is a fine exemplar of the countless books whose explicit purpose was to shape boys during the early twentieth century. On page 246, we even find an unexpected clue as to the reason for this sudden appearance of crafts books for boys. A section entitled "How to Make a Thermometer Back in Etched Copper" is illustrated with a small thermometer fastened to a large copper plate. Forget the thermometer; it is the etched plate that leaps off the page. The decorative pattern, etched onto copper, is a wonderfully pure, clean, and elegant example of art nouveau.

It may seem ironic that the back-to-nature movement of people like John Ruskin and William Morris (described in chapter 9) so rapidly became an expres-

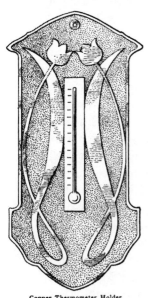

Copper Thermometer Holder

sive language of the new industrial world. But that is what happened. Artisans discovered, for example, that wrought iron made a superb medium for the swirling vines and tendrils that lay at the heart of art nouveau.

The arts and crafts of Ruskin and Morris had, by 1900, become a medium that would serve as the rich and fertile meeting ground between art and technology. The manual arts now became the *industrial arts.*

By the mid-nineteenth century, the industrial world had produced a flood of new domestic products. Looking at the contents of this flood we better understand the arts and crafts movement. It was becoming very clear, as early as the 1860s, that we had created a vast clutter of plaster statues, factory furniture, ready-made clothes, kitchen gadgets, linoleum floors, and low-cost imitation elegance. As these goods poured forth, a public once trained in frugality and circumspection rapidly learned how to be consumers.

With the speed of gossip, sellers invented a new language, a vocabulary of wants. They showed us how to need things we had never even thought about. Author Miles Orvell studies selling in the latter nineteenth century and calls that period *a hieroglyphic world.*[5] Merchants created the new hieroglyphs of taste and of plenty. Stores began displaying great mountains of materiel: pyramids of canned goods, cliff-like arrays of gaudy lamps and parlor statuary. All that abundance now cried out to be bought.

New arbiters of taste rose up to tell us what we should own. Orvell says that a new middle-class norm was created. He calls it "a picturesque eclecticism [that was] Renaissance, Baroque, Classical, and wholly invented. Turned out [in] Grand Rapids, these designs pervaded the households of America."

Since the real thing was generally too expensive, we bought copies. Imitation became a high art. Were we supposed to build in stone? Makers cast concrete to look like stone. They learned to pattern linoleum so it looked like marble or parquet.

The new mail-order catalogs were the primary agents of this new cornucopia. They developed a language just this side of dishonesty. Catalogs showed buyers how to save face when they ordered a $5 imitation of a $25 watch. The new goods could not be sold without first binding buyers to new rules of taste and ownership. Nice people owned statuary. If that statuary had to be stamped out in plaster, so be it.

We must be careful not to label all this embryonic consumerism as pure folly. We throw a thousand inventions up in the air to sample and to savor. We try most of them, but only a handful show us that they deserve a lasting place in our lives. Three-dimensional movies and electric potato peelers have, thus far, gone the way of all flesh; but electric can openers survive.

The sorting process of consumerism is, in fact, the last stage of engineering design. It is a legitimate function that wears the clothing of folly. Yet it blindsided us 150 years ago, and the arts and crafts movement was a needed backlash. We invented the word *homemade* to signify what we felt we were losing.

The arts and crafts movement gained momentum in late-nineteenth-century England, then spread to other countries, reaching its apogee just before World War I. In its purest form, it was retrograde—an attempt to reclaim a medieval ideal. People dreamt of reclaiming a handcrafted life in which all their books, furniture, and clothing were homemade.

What this movement actually did was to create means by which the manufactured world could become connected to the creativity that flows through human fingers. Out of the movement came Frank Lloyd Wright, the Bauhaus School, and new, lean forms of functional design. Out of it came the not-unhealthy idea that the old skills of making things were defining characteristics of manhood.

We can see that mutation occurring with greater focus and sobriety in another book that appeared just before *The Boy Mechanic*. In 1912, Cheshire L. Boone published *The Library of Work and Play: Guide and Index*—a ten-book set on arts and crafts for young people.[6] This was a focused effort to bring the ideals of the movement to the first generation of *Modern*. We may be mechanized, Boone allows, but he insists that, "There was never a time in the history of the world when each race, each nation, each community unit, each family almost, did not possess its craftsmen and artists."

He goes on to identify *individualism* as the commodity that craftsmanship alone will salvage. He also adds an idea that historians were just beginning to understand—that the *true* historical record is to be found in wordless artifacts every bit as much as it is in written documents. (It took another 50 years for that notion to gain wide currency.) Boone also sees, very accurately, how craft serves learning. He says, "The boy makes a kite, a telegraph outfit, or a sled in order to give to his play a vestige of realism. He seeks to mold the physical world to personal desires, as men do. Incidentally he taps the general mass of scientific facts or data and extracts therefrom no small amount of very real, fruitful information."

Boys were generally the sole focus of books like this, but Boone is unusual in that he gives equal attention to boys' and girls' activities. Still, nineteenth-century gender differentiation remains well-defined: Girls do basketry and interior design. Boys build furniture and model airplanes. Gardening could go either way.

The Arts and Crafts ideal of interior design
as shown in *The Library of Work and Play.*

What does come through with almost moral force in Boone's book are two ideals. One is that of individual capability. The other is the ideal of simple and uncluttered design. In that ideal, the arts and crafts people began distancing themselves from many of the ornate excesses of art nouveau. Here we see a kind of segue from art nouveau into art deco. And that ideal still lingered within *my* school curriculum by the 1930s.

We thus find little of retrograde romanticism in Boone's book. Here we are asked to build our own radio or steam engine. Boone strongly emphasizes model-airplane building at a time when the fact of flight had barely reached the general public. For where can one find clean functionalism put forth as explicitly? That is precisely what Le Corbusier (see chapter 11) would say in much more strident terms, 22 years later.

A call to enter the twentieth century with our heads screwed on is what lay at the heart of the arts and crafts movement. Factories and high technology were an obvious fact of life, but they were no reason for giving up hands-on beauty, elegance, and involvement with artifacts. Here that message is sent, with wonderful intensity, to the generation poised to build *Modern.*

One can see a discontinuity, however, between Boone's *Library of Work and Play* and *The Boy Mechanic.* Boone is still expressing reform in nineteenth-century terms. *The Boy Mechanic,* on the other hand, is pure

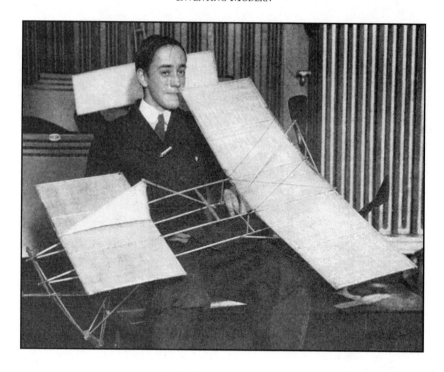

Modern—a book that Le Corbusier would have liked. When Le Corbusier cried, "No door is closed. . . . Everything is relative," he picked up on the same live-for-the-moment sentiment that we read in *The Boy Mechanic.*

The radical new ideas that rode into the early twentieth century took us to a place far beyond the mere collision of industrialization with manual crafts. Those wholly disruptive ideas called for an acceleration of change in every quarter. They swept the arts and crafts movement into regions that would not have made a particle of sense to Ruskin or Morris.

Consider, then, a later book for boys, one that shows how the wind had shifted after World War I. This has the title *The Boy Scientist.* It was written in 1925 by British scientist A. Frederick Collins.[7] The thick red tome has 340 illustrations, no equations, and a writing style meant for high-school boys. The new hyperawareness of genius (described in chapter 4) compels the author to disgorge himself of Einstein before he does anything else.

The first chapter purports to explain space, time, and the fourth dimension; the second tells about matter, force, and motion; and the third (entitled "The Einstein Theory") gets down to the meat of it. The theory is, he tells us, "a subject you ought to know about if you wish to keep abreast of the times." In 1925, everybody was seeking to explain Einstein—dropping balls in moving trains in the vain hope that they would reveal why aging is slowed on a rocket ship to Alpha Centauri. Einstein lay in the grain of *Modern.*

But once Collins dealt with the hurdle of being up-to-date on Einstein, he could get down to his lessons in hands-on, applied science: Subsequent chapters are titled "Astronomy in a Nutshell" and "Chemistry Made Easy," and they deal with photography, crystallography, radio, movies, and flight.

Collins fearlessly explains how to make a spectroscope, a radio, and an X-ray machine. I mention the dangers of those do-it-yourself X-ray machines in chapter 3, and Collins leaves us as unprotected as any of the X-ray pioneers had been. He writes, "Now if you place your hand against the cardboard in front of the fluoroscope, then hold it within 6 or 8 inches of the X-ray tube . . . you will see the bones in your hand . . ."

Do that often enough and you will end up the way Elizabeth Fleischmann did. One section is on "Sulfuric acid and how to make it"; another describes ways to experiment with radium. The book actually offers us more effective means for killing ourselves than *The Boy Mechanic* did, but it now does so under the banner of science.

These new books were, of course, only a variation upon the ancient theme of the boy-as-adventurer. However, that theme gained momentum during the latter nineteenth century. As early as 1853, we catch the sense of it by listening to a Dickens character addressing a young boy in the novel *Bleak House*.[8]

> 'My young friend,' says Chadband, 'it is because you know nothing that you are to us a gem and jewel. For what are you, my young friend? Are you a beast of the field? No. A bird of the air? No. A fish of the sea or river? No. You are a human boy, my young friend. A human boy. O glorious to be a human boy! And why glorious, my young friend? Because you are capable of receiving the lessons of wisdom, because you are capable of profiting by this discourse which I now deliver for your good, because you are not a stick, or a staff, or a stock, or a stone, or a post, or a pillar.'

Then Chadband adds this remarkable and often-quoted couplet:

> 'O running stream of sparkling joy
> To be a soaring human boy!'

Dickens lay a powerful burden of responsibility upon a young boy. If *soaring* meant a buoyant soul to Dickens, the term had mutated 60 years later when *Popular Mechanics* literally advised boys to build their own glider and soar off a cliff in it.

A. E. Housman took a far dourer look at being a boy in 1896. In the loose collection of 63 poems that make up *A Shropshire Lad*,[9] he mourns the death of the lad from Shropshire, sent off to die in one of the wars of British imperialism. This very short poem appears halfway through his lament. Houseman puts these words in the mouth of the lad's ghost:

THEY LEAPED INTO A BOAT.

J. A. Pg. 210

Illustration for George A. Henty's *Jack Archer: A Tale of the Crimea.*[10]

> Here dead lie we because we did not choose
>> To live and shame the land from which we sprung.
> Life, to be sure, is nothing much to lose;
>> But young men think it is, and we were young.

In this chillingly dense and brief poem, Housman sadly says that chancing death is any lad's lot. If his Shropshire lad carried the same responsibility for soaring risk taking that we read in *Bleak House*, Housman at least knew his lad had been mortal.

For several decades before Housman wrote *A Shropshire Lad*, another Englishman, George A. Henty, had been pouring forth books filled with high adventure for young boys. My father read many of the Henty books to me when I was a child, shaping my sense of self. Most were historical novels of derring-do with titles like *With Clive in India, With Lee in Virginia*, and *The Young Carthaginian*. Henty touched at least two generations of boys and taught them history while he fanned the flames of risk-taking.

Boys like me took these stories to heart. Even the doomed Shropshire lad represented a view of ourselves that we wanted to live up to. Behind our risk taking in technology lay our desire to be heroes.

It was all of a piece, and an organization that tapped into the entire skein of it was the Boy Scouts. In 1913, the patriarch of that organization, Daniel Beard, offered a fat book entitled The *American Boy's Handy Book*. It reads rather like a cross between *The Boy Mechanic* and *The Library of Work and Play*. It has the earnest purpose of the arts and crafts movement combined with a good, solid potential for self-destruction.

In his foreword to a 1983 reprint of Beard's book, Noel Perrin grapples with its messages.[11] He warns us to be careful if we try to build the items in the book. He also points out that the boy of Beard's day was regarded as a savage. He characterizes the thinking of 1913 in these words: "A boy of ten or twelve had more in common with wild Indians than he did with his own parents. He probably even had more in common with his dog. Later he would change, of course. He would undergo a spiritual metamorphosis as striking as the physical one his sister went through. From it he would emerge thinking like a man."

Unlikely as that process may sound, it is what often happened. In 1999, a book traced in detail precisely how it worked its way out in one boy's life. Homer Hickam's autobiographical story, *Rocket Boys,* was almost immediately made into the movie *October Sky*.[12] Homer Hickam was fourteen years old and living in Coaltown, West Virginia, in 1957, the year the Russians launched *Sputnik*.

The steel companies were losing interest in Coaltown's deep mining operation. The imminent death of the mine was obvious to everyone but

Homer's father, one of the mine's senior managers. Young Homer sensed the decay of his world on a visceral level. He wanted to become an engineer and build rockets. He wanted to go to work for Werner von Braun.

Football was king in Coalville, a place with no sympathy for foreign rocketry. Homer looked for a book on rocket making and could find none. So he recruited three friends, and they set out to invent their own rocket. Consider the difficulty they faced in building a real rocket that would travel miles straight up.

They needed a chemical propellant and a binding agent for the fuel. They needed to shape the fuel *within* the rocket, and they needed metals to withstand the temperature of burning fuel. The shape of the necessary supersonic nozzle is not only mathematically complex; it also turns out to be an *unexpected* form—a passage that first converges into a small throat section and then diverges in a bell shape.

A true guidance system would have been far too complex. Failing that, they needed to machine accurate tailfins and a launch system to aim it properly. They also wanted means for measuring the height of the flight, for without that any success would be meaningless.

The four boys accomplished all these things, and they did it against a backdrop of economic and domestic chaos, amongst people who could not comprehend their work. In the end, they hurled a rocket six miles into the sky and won a national science fair prize as well. They also threatened significant bodily harm to both themselves and the town of Coaltown along the way.

On the surface, the book leads us through their problem-solving process. But the problems of the rocket fuse with the problems of Coaltown and of the boys' troubled family lives. As the tale unfolds, readers are hardly aware that they are learning college-level thermodynamics, fluid mechanics, chemistry, dynamics, and metallurgy along with those four boys. That's because each of those issues is mirrored in one or another of the crushing social problems that lie in the ever-present background— the problems that go with hacking a living out of a dying company town. It is a remarkable piece of multilevel storytelling.

Homer Hickam went on to become an engineer. He never met von Braun, but he joined NASA in the 1980s, and he eventually did meet the Russian engineers who launched *Sputnik*. In 1997, an astronaut carried one of his old rocket nozzles on the shuttle *Columbia*. Only ghosts of his childhood linger in the remnants of Coaltown, but they are benign ghosts. They are the glorious array of surmountable trials that lay on the way of bringing one of the last soaring boys of *Modern* to his maturity.

Hickam's tale was a holdover of *Modern*, for *Modern* was now being left behind. Perhaps Hickam had to come out of a portion of America separated from the mainstream by the Appalachian Mountains and by poverty. Yet his is a story that was repeated, almost verbatim, by a colleague of mine at the University of Houston, in our chemistry department. His name is Alan (Russ) Geanangel.

Geanangel went to high school in Cadiz, Ohio, also in western Appalachia, and roughly a hundred miles from Coaltown. He and fellow students Bill Steer and John O'Neil developed a rocket, with encouragement from their chemistry teacher, Lucy Patterson. It achieved an altitude just under one mile. That was less than the Rocket Boys managed, but the Cadiz group was at work on a more sophisticated two-stage rocket when Steer's graduation from high school broke up the team.[13]

And so, for half a century, boys saw around them automobiles on the roads, airplanes in the sky, movies in theaters, and books telling them they could replicate all that and more. When I open one of these strange instruction manuals, offering us high adventure as long as we did not kill ourselves, childhood comes back to me in a rush.

My generation really did learn how things worked from books like this. They shaped us; taught us rashness and naiveté; and made it very clear that what one fool can do, another can also do. They were unbelievably intemperate; yet I cannot reread them without grieving the inevitable passing of spring.

The legacy of that para-academic education is shot through our world today, for it spawned invention as nothing else could have done. Once upon a time, we all understood that the game was not to *follow* the instructions they provided, but to *elaborate upon* them. We read those surrealistic instructional manuals and then let our imaginations flow.

And what of the machines we actually produced? Once more *Modern* surprises us. We look next for the special flavor of early-twentieth-century invention.

13

Inventing a
Better Mousetrap

The old saying "Build a better mousetrap, and the world will beat a path to your door" is widely attributed to Emerson. In his article on the history of the mousetrap as an invention, writer Jack Hope points out that Emerson really wrote something quite different: "If a man has good corn or wood, or boards, or pigs, to sell, or can make better chairs or knives, crucibles or church organs, than anybody else, you will find a broad hard-beaten road to his house, though it be in the woods."[1] Emerson said nothing about mousetraps. In 1889, however— (seven years after Emerson died)—someone quoted him as having said, "If a man can write a better book, preach a better sermon, or make a better mousetrap than his neighbor . . ." and so on.

Emerson had merely been pointing out that quality prevails in the marketplace. Yet something remarkable occurred when his remark touched, and adhered to, the mousetrap. At that point a metaphor of compelling power was created to embody our new view of invention.

We have, ever since, made a vast investment of ingenuity in mousetraps. By 1996, the U.S. Patent Office had issued over 4,400 mousetrap patents, and some four hundred people per year were continuing to seek them. Since only 20 or so of those patents have ever made money, the impetus for all this *mousetrappery* had to lie in something other than mere gain.

The mousetrap problem was actually given its present-day solution in 1899 (only ten years after the quasi-Emerson quotation became common currency). One John Mast of Lititz, Pennsylvania, filed for a patent on his now-familiar snap trap—the one with heavy spring-steel wire that swings down and breaks the mouse's neck when it nibbles cheese on the trigger mechanism. Now, over a century later, there has yet to be a significant, lasting improvement upon Mast's invention.

The snap trap, patented in 1899 and yet to be surpassed by any other mousetrap. (Photo by John Lienhard)

Ever since Mast, the inventive muse (or mouse?) has nevertheless kept generating mousetrap patents. Inventors have cooked up an unending series of gadgets for mashing, cutting, and maiming mice—for drowning them or for catching them alive. Early in the twentieth century, people tried electrocution. The problem with it was that an electrocuted mouse continues to fry until someone smells the mess.

Esthetics and mercy have been the twin factors that absolutely determine what the public will and will not use. In the 1980s, a superglue trap came out. It worked, but homeowners found themselves faced with a screaming mouse, still living but dying of exhaustion, glued to a piece of sticky cardboard.

If mice are to be killed, most people can deal with a quickly broken neck; the more gruesome stuff will not sell. When snap trap makers found most people throwing the trap out with the mouse, not even trying to disengage it, they followed the public's lead and began advertising snap traps as "disposable." The public makes it clear how far it is willing to go in the grisly business of holding a competing species at bay.

As we look at all this human energy, we realize that building a better mousetrap has had rather little to do with finding better means for killing mice. (If the intent *was* only to improve mousetraps, here is the rare instance in human history when a goal has been addressed with such energy and not been achieved.) Something else was at play here. Something more elusive than function turned the lowly mousetrap into the embodiment of Emerson's remark about economics.

Mousetraps appear to lie well within the grasp of the *everyman* inventor. Testing a new concept for catching mice is an expression of personal art and craft. The very smallness of the "better mousetrap" appears to give anyone access to the genius of invention. Thus the kitschy metaphor of

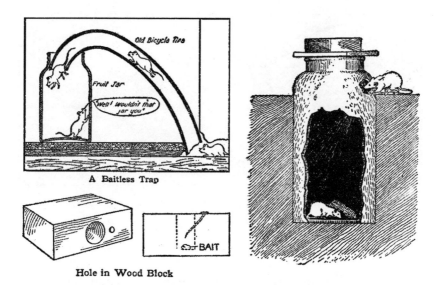

Three of the mousetrap ideas offered in *The Boy Mechanic*, 1913.

the mousetrap was a place where arts and crafts could seemingly converge with the hidden genius that we all felt rumbling within us.

The new books for boys picked up the theme. *The Boy Mechanic* included six kinds of mousetraps that the "boy-savages" of the early twentieth century could build or further develop.[2]

If building a better mousetrap became the metaphor for the new inventive energy, this metaphor had to be acted out. The number of pat-

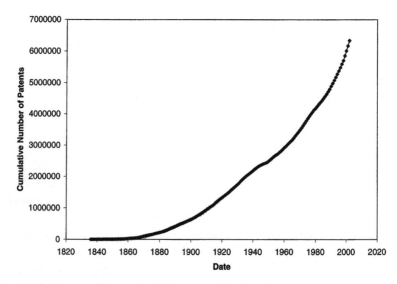

Graph of the accumulation of U.S. patent numbers, based upon Patent Office statistics.

ents in general began rocketing upward just about the time the "better mousetrap" phrase came into being, and mousetrap patents rose alongside the more serious ones. It would almost seem that a fraction of each year's patents had to be offered up to the icon—that the metaphor had to be honored. The continuing rise of mousetrap patents, long after the snap trap was invented, seems to have functioned rather like a tithe—an almost sacramental expression of our belief in invention.

Since the arts and crafts movement came out of Great Britain, it is useful to look at a book assembled in 1979 by Rodney Dale and Joan Gray.[3] It displays six hundred of the 140,000 British patents registered between 1901 and 1905. Reading it, we gain a hint of the way the movement was beginning to express itself in invention.

A few of the patents in the book are ones that changed history, but only a few. The English patent for a Wright brothers glider is among the more significant ones in the collection. In many other cases it is entirely clear that the patented devices would never have functioned if they had been made. Many are obvious poppycock. We find a wonderfully unlikely flying machine intended to get off the ground by flapping its inadequate wings like a kiwi bird. Only a few of the many helicopter and dirigible patents could conceivably have been made to work.

A surprising number of the inventions were meant to serve the popular Edwardian theatre—artificial horses, magic tricks, and means for seeing around ladies' hats. We also find a shocking array of straps, stays, and heavy metal meant to shape the female body into forms meant to satisfy fashion by defying nature.

Particularly poignant is the trail of false starts that finally resulted in the set of household appliances we actually use today. Gadgetry of this kind has formed a steady parade through the patent offices for two centuries now. Here we find a specialized orange peeler, an anti-bed-wetting device, a mechanism for picking up pins, a special pickle spoon, and an apparatus meant to help people swallow pills. These were the broader expression of the American mousetraps, I suppose—the arts and crafts of daily problem-solving.

Not unexpected is the usual profusion of impossible perpetual-motion machines. Most are just variations on the so-called overcentered wheel, first suggested in India during the twelfth century. One of these devices is new, and not, strictly speaking, a perpetual-motion machine. It is a cable that would reach 150 miles into the sky and take electricity out of the ether. Installing this 80-ton cable and holding it up poses an obvious problem, so the inventor explains that the ether will hold it in place, once one has found a way to make it stand up.

The meat in this Dagwood sandwich includes Fleming's patent for the first radio diode, as well as primary patents for puffed wheat, nitric acid production, gyro stabilization, and embryonic forms of both the hovercraft and radar.

British patents were good for three years. Then one had to pay a fee to renew them. Only half of the inventors took the trouble to do so, and a scant 2.5 percent of patents were still hanging on after 13 years. There is an oddly frivolous quality in all this patenting, as a new mania for inventing things became one of the characteristics of *Modern*.

This collection of Edwardian patents illustrates the implicit message that we all listened to as we read the new how-to-do-it books for young people. They reminded us that we had to be ready to fail 50 times if we wanted to succeed once. They made us know that we might have 50 good ideas in our heads, and, if they all failed, we could still have 50 more. It is, in that sense, far less a record of failure than it is a record of the mind-stretching fun we were now having with invention.

And so we set out to be inventors, not quite knowing what it was we wanted to invent. In 1899, the same year that Mast invented the classical snap-trap mousetrap, another invention, even more minimalist, and far more universal, appeared. William Middlebrook patented a machine that would make wire paperclips. As Henry Petroski explains, he did not patent the paperclip itself.[4] However, in one corner of Middlebrook's patent drawing is the clip that his machine was intended to make.

This paperclip had the round top and bottom so familiar today. We call it the *Gem* paperclip because Middlebrook developed his machine for the Gem Company, in England. And where did his paperclip design come from? Petroski finds evidence that it had already been in use in England.

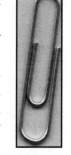

Yet, like the mousetrap, paperclip patents continue to flow forth. Some years ago, I sent a paper off to an editor. I happened to put it together with a paperclip that I had carried back from Yugoslavia. He was delighted, since he collected paperclips, and this was one he hadn't seen. After that, I found a different clip for the material I was sending him every few weeks, and I never ran out of variations.

Until the twentieth century, people had bundled papers with ribbon or string. If the sheaf was small enough, a straight pin or a kind of clothespin was used to hold papers together. When Petroski studied this problem he found a great deal of late-nineteenth-century experimentation, but very little in the way of patents. In 1887, Philadelphia inventor Ethelbert Middleton patented a soft piece of metal that could be bent into place to hold paper, and, in 1898, an American named Matthew Schooley patented several paperclips that had much in common with the Gem clip.

The paperclip patent often *credited* with being first is that of the Norwegian Johan Vaaler. He devised his clip in 1899 (the same year as Middlebrook's paperclip-manufacturing machine); however, Norway had no patent office. Vaaler had to file an American patent for a set of square and triangular clips. He received his patent in 1901, and, while clips similar to his may be found today, Middlebrook's Gem design remains as definitive among paperclips as the snap trap is among mousetraps. It is the standard to which we always return.

If you pick up a modern Gem paperclip and study it closely, you will discover that (like the snap trap) it is a wonderful compromise solution to a very hard problem. Consider the pitfalls awaiting a new paperclip: It should exert a clinging grip; it should not tangle with other clips in a box; it should be easy to apply and to remove; it should not tear the paper or leave rust marks; it should be cheap and easy to make; and its use should be obvious.

One Gem clip maker said he got ten letters a month suggesting improvements. But people could improve one area only at the expense of another. So the paperclip remains a tantalizing exercise in elegance and sophistication. Like the snap trap, it is a compromise whose lingering challenge still teases us today.

The huge majority of paperclip and mousetrap patents thus were awarded *after* the definitive mousetrap and the definitive paperclip had been created. But then, *Modern* would be the epoch of the patent. And patents have, in some ways, become a blight. Take the case of Niels Christensen, who was, according to George Wise, harsh and uncompromising; and who made his finest invention when he was over 70.[5]

I encountered Christensen's invention in 1951, just after I had left engineering school to join the Boeing Company. Early in the job, I had to design a valve. The valve stem had to slide inside a cylinder and seal in hydraulic fluid. An older engineer saw me trying to select a gasket and said, "That's no way to do it. Use an O-ring." "What's an O-ring?" I asked.

O-rings had just come into use during World War II. I had never seen one in school or in any of my own manual work. An O-ring is a skinny doughnut-shaped piece of rubber or neoprene that is fitted into a circular slot with a square cross-section. The way it works is subtle:

As the ring mates with another surface, it distorts in such a way as to drag in a film of lubricant from the moving shaft. It also snugs down into one side of the slot and tightly seals against leakage. Those little rings have truly revolutionized sealing. They are almost as common as paperclips today and far more common than mousetraps. Once we've learned what goes on *inside* the machines around us, it becomes hard to imagine life without them.

Their inventor, Christensen, was born in Denmark in 1865. He studied as a machinist, and he came to America in 1891. During the 1890s he invented a fine new air brake for electric streetcars and was soon up to his neck in litigation. His patent rights passed from company to company until Westinghouse owned them.

Christensen did not suffer perceived injustice lightly. He fought through the courts while he went on inventing new gasoline engines, car and airplane starters, and more braking systems. He began looking at hydraulic-system seals in 1933. It took four years of experimenting with rubber rings in slots to get it right. Christensen filed his first O-ring patent in 1937, when he was 72 years old.

In 1941, he licensed the patent to United Aircraft and was set to get rich. Then, after Pearl Harbor, the U.S. government bought out all the key military patents and gave them away to manufacturers. Christensen received a lump payment of $75,000, and O-rings belonged to the government. He mounted his last great court battle, which ended 19 years after his death. In 1971, Christensen's heirs managed to collect $100,000 for his work.

O-rings.

Here I must insert a personal note: I loathe these patent-battle stories, for they so demean the creative process. I want to remember Christensen for giving us one of the great inventions when he was 72—for changing the world after the world had begun to call him old. Surely that was better payment than ever came to any of the people who profited from his ideas. But *Modern* was now the time of the patent—a time when all we boy mechanics meant to get rich from our better mousetraps.

The elements we have been considering in chapters 12 and 13 strongly converge in one particular inventor, Edwin Howard Armstrong.[6] Born in 1890, he was 15 when he read *The Boy's Book of Inventions*. Straightaway, he declared that he would be an inventor of radios. Radio was then younger than he was, and he dove right in. Seldom was a decision to do something worthwhile with one's life ever as clear-cut.

By 1912, Armstrong was a student at Columbia University spending a huge amount of time doing research in wireless telegraphy and radio. But it was on a hiking trip in Vermont that he suddenly saw how he might regenerate an antenna signal. He had found out how to so enhance a radio signal that it was possible to do away with earphones. The next year, his professor gathered engineers of the American Marconi Company at Columbia, and Armstrong demonstrated his *regenerative feedback circuit*. Armstrong showed them how he could pick up signals from Ireland by rewiring a standard radio tube.

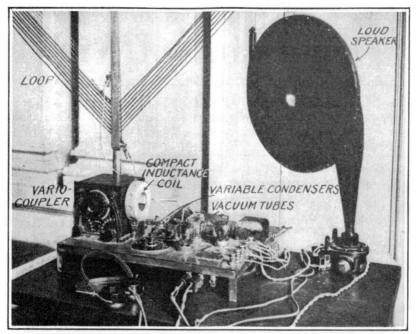

General arrangement of standard radio instruments and loop for the Armstrong super-regenerative receiver, using a loud-speaker

(from the September 1922 *Scientific American* magazine.)

Seven years later, *Scientific American* celebrated Armstrong's next variation on the idea, his new super-regenerative circuit, as a milestone in radio.[7] The magazine noted that his regenerative circuit was now in wide use, and it warned amateurs that this new version was very complex. It was not the sort of thing that others could easily reproduce in their workshops.

Armstrong was big and athletic. He played tennis with hungry intensity, raced automobiles, climbed mountains, and (when he had no mountain at hand) climbed radio towers. As a young major in the World War I Army Air Service, he worked on the development of aerial radio communications.

While he worked, however, he also wondered if he might track an enemy airplane by reading the weak, high-frequency electric signal generated by its ignition equipment. That set the stage for a major conceptual leap. He saw that he could mix an incoming radio signal with an imposed signal and produce a new signal with a frequency that was easier to amplify.

He called the arrangement his *superheterodyne* system, and he went on to spin it into FM radio. By 1933 he was able to show a practical FM system to his old friend David Sarnoff, president of RCA. Sarnoff experimented with FM but eventually abandoned it because it meant too radical a change in the radio infrastructure. Armstrong finally launched his own FM system and did not fare well in that business. The technology, on the other hand, caught on. After World War II Armstrong fell into a quicksand of lawsuits with RCA over FM patent rights.

Armstrong had met his wife Marion MacInnis in 1923, and he had wooed her in odd ways. He drove her around in fast cars. He scaled the 450-foot RCA transmitter tower for her. He gave her the first portable superheterodyne radio for a wedding present. Their 30-year marriage was marked by tension as his love for her and his obsession with work collided.

By the end of World War II, Armstrong was swimming in losing legal battles, and his youthful verve was caving in. Finally, after having Thanksgiving dinner with friends in 1953, he confessed his rising desperation to Marion. She suggested that it might be time for him to retire, and he flew into a rage. He swung a poker at her and struck her arm, and she moved out.

From that point on, Armstrong went downhill very rapidly. Finally, on January 31, 1954, he gave up. He wrote Marion a wrenching letter apologizing for what he had done and what he was about to do. Then that lifelong lover of heights removed the air conditioner from the window of their thirteenth-floor apartment and leapt to his death on a roof ten stories below. His wife had greater staying power than he did. She won the final patent victory in 1967.

Our now-ubiquitous FM radios came into being as the result of Armstrong's teenage certainty that he could scale the mountain, win the prize, and give the world something good and new. It is a frustrating story of a journey from the wonderful verve of *The Boy's Book of Inventions* all the way to Greek tragedy, and it brings Dickens's words back to haunt us:

> O running stream of sparkling joy
> To be a soaring human boy!

Yet that sentiment cannot be forgotten, for it underlay so much of the texture of early-twentieth-century invention. Like flight and other new technologies, invention itself was a door into freedom, and it was an expression of freedom. We find a fine example of that underlying an article in a 1914 New Orleans newspaper.[8]

The article reported an advertising demonstration in which an inky, thick smoke had been created. Then a costumed Indian assistant, Big Chief Mason, had donned the new Morgan safety hood and spent 20 minutes in the smoke without ill effect.

What neither the onlookers nor the writer of the article knew was that "Big Chief Mason" was, in fact, Garrett Augustus Morgan, the safety hood's inventor. Morgan's mother was a freed slave, and since Morgan's birth in 1877, American racism had steadily worsened. The only way he could sell the hood was to hide his black identity. So he demonstrated it in a sort of cigar-store-Indian disguise.

Fortune found Morgan out in 1916 while he was working in Cleveland. Workers were drilling a new tunnel under Lake Erie to supply fresh water. When they hit a pocket of natural gas, it exploded. Workers were trapped, suffocating in smoke and dust. Someone had heard about Morgan's safety hood and called him in to help. He and his brother suited up and repeatedly went into the tunnel. They saved two lives and recovered four bodies before officials closed the tunnel to rescue efforts.

Unfortunately, Morgan's race was now exposed, and that hurt his sales. Still, the gas masks used in World War I were kin to Morgan's safety hood. And that was only one of Morgan's inventions. In 1923 he came up with the device that led to the three-way traffic light.

He saw that existing mechanical stop-and-go signals were dangerous because they had no caution signal to buffer traffic flow. So he patented a three-armed signal that would indicate stop and go for traffic in two directions. It also had a four-way stop for pedestrians and a signal for moving forward with caution—the forerunner of today's yellow light. General Electric bought his patent for $40,000, which was a huge sum back then.

By the age of 30, Morgan had set up a small sewing-machine shop, where he patented sewing-machine improvements. Sewing-machine needles functioned better when they were polished, and Morgan found, by chance, that commercial liquid polish would also straighten hair. He figured out how to make the stuff into a cream. He created his own company and marketed it.

When I was a child, many African Americans straightened their hair so as to better blend into a white landscape. All of Morgan's other inventions saved lives and accommodated human need. Hair straighteners met another kind of human need in 1910, and Morgan did what he had to do in the worst of times.

And so Americans built their various mousetraps. The large inventions, the earthshaking and society-altering engines of human ingenuity, were coming into being. But, as they did, countless "boy mechanics" (some of whom inevitably were female) saw to the texture of daily life in America. The better mousetraps of the early twentieth century were close-coupled with that risk-taking verve.

The young Charles Lindbergh was one of those boy mechanics, and not enough is said about how his daring was manifested in invention.

Lindbergh did what others had not been able to do because he knew machines. He played a large role in designing the highly specialized *Spirit of St. Louis* in which he flew to France.

The medical work Lindbergh did a few years later has been largely forgotten, but it was highly significant. In 1930 his sister-in-law suffered heart trouble. Doctors could not operate without stopping her heart, and that would have been fatal. Since that struck Lindbergh as a solvable problem, he talked to Alexis Carrel, who held the Nobel Prize for his work in organ transplants and suturing blood vessels.

Carrel was respected, but he was odd. His operating room and his scrub linens were solid black. Author Christopher Hallowell describes how Carrel "flirted with arcane mysticism" and harbored bizarre racial theories.[9] But then Lindbergh himself was an odd duck, alienating people before the war with his isolationist ideas. The two took a real shine to each other.

Carrel was already asking if an external blood pump might not be made to sustain the body while he operated on the heart. Lindbergh studied the problem and quietly went off to the Princeton University glassblower. Two weeks later he came back with his own blood pump. Carrel was delighted and invited Lindbergh to continue work in his labo-

ratory. Lindbergh did. At first he produced a series of pumps that did not quite work.

In 1935 (soon after his son had been kidnapped and murdered) he produced a working blood pump that would sustain organs outside the body during transplantation. It became known as the *glass heart*. Lindbergh also produced a good deal of the supporting technology. He made a centrifuge to separate blood plasma without damaging it.

Carrel sang the praises of the work. He and Lindbergh appeared together on the cover of a 1938 *Time* magazine, admiring their pump. The press wrote about transplants and implants and about the medical miracles right around the corner. Maybe the pump itself could be miniaturized and used to replace the human heart.

Then came World War II: Lindbergh and Carrel both responded to new and more urgent priorities. Most of the pumps were broken up for the platinum in them. Carrel himself died of heart failure during the war, while Lindbergh flew combat missions in the Pacific.

Lindbergh's perfusion pump for circulating blood. (Image courtesy of the National Museum of American History, Smithsonian Institution)

Today the artificial heart is a reality, but it embodies a technology that was jump-started by one more soaring boy mechanic, one more builder of better mousetraps. *Modern* was not just the age of the patent; it was the age of a special—even mythical—new creature. It was the age of the *inventor*, a word that once carried connotations unlike any that attend it today.

That began to shift after World War II. I think we can read the shift in two great inventors who were shaped during the Great Depression and who became fully defined during the war. The first was Chester Carlson, whose story I shall pick up at its end: We join Carlson one afternoon in 1968 when he had a few hours between meetings.

Carlson was then 62, a multimillionaire who did not make friends easily and who lived inside his own head. Carlson went to a movie that afternoon and, as he watched the images flickering on the screen, died of a heart attack.[10] That lonely death was curiously appropriate.

Images and loneliness had been the twin themes of Carlson's life. His father died when he was 13, and after that he had to work both to help support his desperately poor family and to put himself through school. In one job he served as a printer's assistant, and he began thinking about the difficulties of reproducing images.

He labored on like a galley slave, all the time dogged by a sense of inferiority and inadequacy. He graduated from the California Institute of Technology during the Depression, and then barely stayed employed while poverty swirled about him. In his off hours he worked in a chemistry lab he had created in his kitchen. He sought to invent a cheap way to reproduce images on paper using electrostatic means.

Carlson patented a copying process in 1937, before he had really figured out how to make it work. He hired a German refugee named Otto Kornei to help him. Working on a budget of $10 a month, they finally managed to reproduce an inked message by electrostatic means. Kornei saw little future in the process, so he went on to a regular job. Carlson spent the next six years looking for corporate backing.

The Battelle organization finally bought into his patent, and Carlson vanished into the work of developing the process. First his marriage fell apart. Then Battelle gave up on the process. Finally, a little company called Haloid bought the patent rights and hired Carlson.

Haloid turned to a Greek scholar for help in naming the process. Since it used no photographic liquids, he suggested that they base the name on the Greek word for *dry—xeros*. He suggested that they call it "Xerography." That word was simplified to "XeroX," and Carlson's dream was finally on its way. It took another 13 years, until 1956, to produce the first really successful XeroX machine, but when it did, Carlson became wealthy to the tune of $150 million.

He spent his last few years working as hard at giving his money away as he'd worked to earn it in the first place. He made sure that his old colleague Kornei also became wealthy. The money was not really part of the deal that Carlson had forged with his life. He had struggled to survive and to gain self-acceptance, he had chased the images that danced in his head, and, in the end, he had given the world much more than it had given him. Perhaps those of us old enough to still use the word *XeroX* for a photocopy should learn to call it a *Carlson* instead.

But Carlson was driven by new forces—forces that lay beyond *Modern*— and his education had come out of college more than out of those old books for boys. The word *inventor* was ceasing to carry imagery of the soaring boy mechanic / mousetrap builder.

Today, we occasionally try to resurrect that archetype (remember the movie *Back to the Future*). But when we do, we have to explain ourselves as we go along. The word *inventor* is no longer a shorthand that brings up a completed character in the minds of readers and viewers.

Today's inventor is a college-educated corporate employee and team player. Not for nothing was Lindbergh called *the lone eagle*. The very word *invention* has given way to gentler, less threatening, terms, such as *innovation* or *research and development*. Patents have become certificates of corporate ownership, no longer scalps on the belt of some lone warrior.

Let us then meet one last transitional figure, one major inventor who was clearly more a creature of the world today than that of the early twentieth century. This is an inventor whose work, like Christensen's, I encountered not long after I finished my bachelor's degree in engineering.

By 1957, I was a teaching assistant helping undergraduate students to run a lab experiment at the University of California at Berkeley. We measured stress distribution in a loaded model made from clear plastic. We did that by shining polarized light through the plate—that is, light with only one direction of vibration. Since the index of refraction of the plastic varied with stress, the passage of light was affected. After the distorted rays emerged from the stressed plastic, they passed through another sheet of unstressed Polaroid plastic. That plate blocked light, or allowed it to pass, depending on how stress had affected the light where it had passed through the plastic model.

The result was a sort of Rorschach picture in colored bands. The bands made a contour map of stresses in the plate, and they offered a shortcut around mathematics that would become reasonably tractable only after we had digital computers.

Polarization was known 180 years ago, but more as a scientific curiosity than as a working tool. Then around 1930 several things happened. A woman named Helen Maislen studied physics at Smith College, where one of her professors coined the term *polaroid*. Soon she married Edwin

Polaroid Stress Analysis   Computer Stress Analysis

A plate with a slit running halfway across from the left is subjected to a vertical tensile stress. Both Polaroid and computer analyses shows how stresses are intensified at the end of the slit. (Images provided by Professor Krishnaswamy Ravi-Chandar, Department of Aerospace Engineering and Engineering Mechanics, University of Texas)

Land, a young physics student who had been turned on by courses at Norwich College. Land went on to Harvard, but he became so engrossed with polarization that he dropped out.[11]

In 1937, Land formed a company to produce a new polarizing plastic. Of course he named it *Polaroid*. By 1943 he had become a 34-year-old business wunderkind. That year found him on vacation with his family in Santa Fe, taking snapshots. Then his three-year-old daughter complained, "Why do we have to wait so long to see the pictures?"

Why indeed? That afternoon Land walked through the old town chewing on the question. We shouldn't have to wait, he thought to himself. During that walk, he invented the Polaroid camera in his head.

The camera was based on a new film that developed and printed immediately. At first Land got sepia images; by 1950 he had a black-and-white system. He had invented a color Polaroid camera by 1959, and it was on the market in 1963. Of course, these Polaroid cameras had nothing to do with polarized light. Land had now taken an entirely different direction.

By the late 1960s, half the households in America had Polaroid cameras of one sort or another. When Land retired in 1982 he was a 73-year-old billionaire with 533 patents to his name. Like Armstrong, Land was a reticent and driven man. He once said, "Anything worth doing is worth doing to excess."

That upsets our ideas about moderation and balance. Yet it also recalls an old Biblical idea: "Because you are lukewarm—neither hot nor cold—I will spew you from my mouth." Well, there was nothing luke-warm about the way Land threw himself at invention. And that much reflects the intensity to which we were all asked to aspire in the early twentieth century. Perhaps invention will always show us that face. Nothing that lies as close to the human heart as invention will ever be done with tempered circumspection.

At the same time we sense, in Land, a transition from the mentality of Christensen, Morgan, Lindbergh, and Carlson, to something more sedate. The fine madness, the ecstasy and the agony, of invention are somehow attenuated. He was a superb inventor, but no longer a full-time dweller in the *neverland* of *Modern*.

Both Carlson and Land reflect the gathering influence of the Great Depression and of war. I suspect that the special mentality of *Modern* may actually have been fed by World War I, but that mentality was eventually destined to die out under the impact of a second, even more ghastly, war.

We must therefore ask what happened when the "savage boy inventor" joined with war, as it had been conceived in the nineteenth-century. Within the rising tide of *Modern*, those two concepts combined with frightening volatility. World War I, for all its horrors, was not enough to burn the reaction out. We next look at how this particular fusion process took place, and what eventually became of it.

14

War

I t was early in the morning of August 10, 1915. Thirty thousand fresh Turkish troops came down off Sari Bair hill on the Gallipoli Peninsula. They fell upon the tired, badly positioned forces of the British and their allies. Twelve thousand Allied troops perished in furious hand-to-hand combat. While an even larger number of Turkish soldiers died that day, the Turks nonetheless won, and Allied hopes for moving on to secure the Dardanelles Straits were ended.

Among the dead lay a 27-year-old British lieutenant named Harry Moseley, the most promising physicist of his age. Even the German press, the enemy press, lamented his death. "Ein schwerer Verlust," they wrote: "A heavy loss"—for science.[1]

Moseley had worked with that large, loud, much-loved, and foul-mouthed genius from New Zealand, Ernest Rutherford. Rutherford had already created the modern theory of radioactivity. He won the Nobel Prize in 1908, but greater work lay ahead. Rutherford and his collaborators sorted out atomic structure. In 1912, after working with Rutherford, the Danish physicist Niels Bohr wrote the quantum theory of electron orbits.

Moseley was still only 23 when he joined Rutherford at Cambridge University in 1910. Rutherford had put him to work studying the radioactive emissions from atomic nuclei. Moseley was a patrician and as serious as an undertaker. He worked day and night, he openly disapproved of Rutherford's language, he went his own way, and he tore into the problem of finding the electric charge in an atomic nucleus.

He invented elegant and simple means for bombarding samples with cathode rays. He photographed the resulting X-rays and then made the bold theoretical leap that allowed him to identify an atom by the charge on its nucleus. Within days, Moseley had set the basis for the periodic table. He did much more, but that was his greatest contribution.

As World War I began, his widowed mother remarried. In a letter to his sister he expressed doubt. His mother was gambling "a modicum of happiness . . . on the excellent [but uncertain] chance of getting more." That was Moseley—stiff, serious, analytical, and upright to a fault!

Moseley joined the army soon after the first shots of the war were fired. When I talked about Moseley on the radio some years ago, a Houston chemist, Dr. Edith Sherwood, contacted me to say that her chemist father had been Moseley's contemporary. He had told her that Moseley had enlisted because someone had mailed him a white feather—the traditional accusation of cowardice. Patriotism and duty were bred in his bones, and that white feather must have been like acid in a wound.

Back in the late 1980s, I heard about a smart 16-year-old student who won the Westinghouse Science Talent Search. He had used X-rays to find pollutants in clamshells—a pure variant on Moseley's technique for analyzing nuclei. When someone mentioned Moseley to him, the puzzled student replied, "Who's Moseley?" Sadly, that is a question most people ask when they hear the name of this fallen foot soldier who helped to shape the periodic table.

However, if Moseley's work in physics was modern, Moseley himself was very much a creature of the nineteenth century. I offer him in contrast to an American who, although he was 38 years older, clearly demonstrated the freewheeling mentality of *Modern* that was so absent in Moseley.

He was Hiram Maxim, inventor of the Maxim machine gun, which may have been the very instrument of Moseley's death. We catch Maxim's distinctive flavor when, toward the end of the nineteenth century, he presented himself to the police in St. Petersburg, Russia. He was there to sell his Maxim gun to the czar. He reports the conversation with the official at some length in his autobiography, *My Life*. I won't quote it all verbatim; however the gist of it went something like this.[2]

"Your name is Jewish; all your family names are Jewish. Are you a Jew?"

"No."

"What religion have you?"

"None whatever—I never had any use for one. My people were puritans."

"Well, no one can stay in Russia without a religion!"

"Very well, put me down as a Protestant."

"And that," Maxim tells us, "is how I became a Protestant." Now a Protestant, Maxim went on to become a Sunday-school teacher. He tells about it thus: "Having become a Protestant, I essayed to write a Sunday school lecture, taking for my text the Gadarene swine story. This falling into the hands of George A. Stevens, the artist, he sent me [a] sketch that I might see myself as others saw me."

Maxim's somewhat weird sense of auto-parody was one of his life-long distinguishing traits, and it helps to explain what made him tick. He provides a disarming what-you-see-is-what-you-get self-portrait—a picture of a man who is uncommonly comfortable within his own skin and suffering from very little need for self-justification.

The two Maxim brothers, Hiram and Hudson, were born in Maine in 1840 and 1853, respectively. Both became noted inventors. Hudson made his name in manufacturing explosives, but older brother Hiram was, by far, the more prolific inventor. Early in his long career, he worked with gas illumination. He went on from that to develop electric lighting systems, even before Edison.

(From *My Life*, by Hiram Maxim, 1915)

(From *My Life*, by Hiram Maxim, 1915)

Then, in 1883, a friend told him, "Hang your electricity. If you want to make your fortune, invent something to help these fool Europeans kill each other more quickly!" Maxim took that advice. By 1885 he had invented the first single-barrel machine gun. The Maxim gun fired 666 rounds a minute, and it changed warfare. He sold vast numbers of his guns to Russia, which soon took them to war with Japan. Maxim proudly tells us that "more than half the Japanese killed in the late war were killed with the little Maxim gun."

The Russo-Japanese war was a storm warning of the slaughter that would begin a decade later. Maxim became a British citizen, and as the Maxim guns (and nastier guns that followed) made his name, he gained an English knighthood, and an entrée into the aristocracy and the royalty.

Maxim's photograph of his flying machine taken on its launch rails in 1893.

Yet guns were only another waystation for a very fertile mind. In 1891, 12 years before the Wright brothers flew, Maxim began work on a huge flying machine, which he built at a cost of £20,000. He describes the effort in the 1895 *Century Magazine*.[3] His huge, multi-winged aeroplane had a one-hundred-foot wingspan, and it was driven by two relatively lightweight, 180-horsepower steam engines of his own design.

He began flight tests in 1894. On the third try he powered the plane up to 40 miles per hour, whereupon it left its launching track, flew about two hundred feet, and crashed. After that, Maxim cut his losses and went on to other inventions.

And yet, the intensity of commitment that Maxim put into the project *before* he walked away from it is astonishing. By 1891, he had already written a *Century Magazine* article in which he described his elaborate preliminary testing of the engine for his flying machine.[4] The article began with a critical look at French attempts to build powered balloons and concluded that they are a clumsy form of flight. A bird, he pointed out, weighs six hundred times more than the air it displaces. Yet (for example) a goose in flight never exerts more than a tenth of a horsepower.

He concluded that heavier-than-air flight was the way to go if we meant to fly. He also recognized that, although birds combine lift and propulsion in the wing, that combination is too subtle for us to mimic. Maxim knew it would be necessary to separate the wing from the propeller.

To accomplish that end experimentally, he built a central tower with a 32-foot rotating arm to measure the effectiveness of propellers and wing surfaces. At the end of the steam-engine-driven arm was a propeller with a streamlined engine pod and a short section of a wing. That test configuration circled the tower at speeds up to 60 miles per hour, while an electric motor inside the pod drove the propeller.

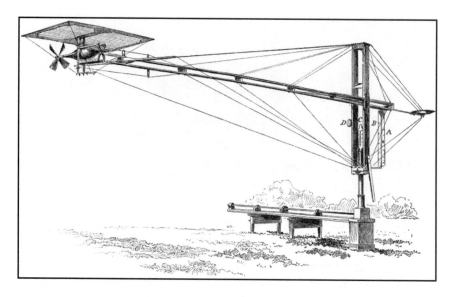

Maxim's drawing of his "Machine for testing the efficiency of the screw-propeller and the lifting power of the aeroplanes."

The apparatus offered means for measuring power inputs to both the propeller and the rotating arm. Maxim's instruments allowed him to separate out the forces of lift, thrust, and drag. His energy inventory revealed that, at 60 miles per hour, the propeller might use 16 horsepower to lift the wing, and another 35 horsepower to overcome drag, as well as its own inefficiency.

Even with such fine preliminary work, Maxim's flying machine failed. Maxim had done a superb job of learning how power is spent during flight, but he had not solved the crucial problem of controlling a moving aeroplane. Later, he said that he wished he'd had one of the new internal-combustion engines and had built a smaller plane. The Wright brothers, of course, did just that, as well as solving the control problem, 12 years later.

Two features of these articles are very revealing. First, Maxim openly shows his readers photos of his crashed aeroplane. There is no dissembling here, no sugarcoating. The machine crashed, and that is a part of the story. He shows no more defensiveness about the failure than he does later about anything else.

The other feature is the way in which these articles reveal the purpose that Maxim sees for his aeroplane. Wilbur Wright had bluntly said that *sport* would be the primary purpose of flight. Maxim expresses a very different intention: The last sentence in his 1891 article comes directly back to war. He writes: "Flying-machines will therefore be employed only by the rich and highly civilized nations. Small nations and half-civilized tribes will still have to content themselves with their present mode of warfare."

The concluding sentence of his 1895 article drives that point home even more strongly: "If one half the money, the time, and the talent which has been employed by the French balloon corps in their fruitless attempts to construct a navigable balloon should now be employed in the right direction, the whole question of aerial navigation would soon be so perfected that flying-machines would be as common as torpedo-boats, and the whole system of modern warfare would be completely changed."

It may seem that Maxim (along with the generals and princes he ran with) saw nothing more soul-reaching in war than he did in religion. However, throughout Maxim's writings, we find a black cloud of uncanny predictive accuracy. We wish we could laugh him off; but we cannot.

In 1889, Maxim and another maker of armaments crossed paths. That year, he and his contemporary, Alfred Nobel, sued the British Government for infringing their patents. Each had developed a smokeless explosive powder and submitted it to the Explosives Committee of the British War Office. The British had looked over both formulae and then created their own version, which became known as *Cordite*.[5]

If Maxim's role as an armorer was unashamed and unambiguous, Nobel presents a far more complex face. He constantly struggled to erect an acceptable rationale around himself and his work. Nobel was born in Stockholm in 1833, the same year his father went bankrupt. The family moved from country to country while his father tried different businesses.[6]

By 1861 his father was producing the terribly unstable new explosive, *nitroglycerin*. American railroads were using it to blast out track beds, and Europeans were using it to tunnel under the Alps. Nitroglycerin killed workmen with tedious regularity. In 1864, Nobel had his youngest brother Emil working on the purification of nitroglycerin at a small plant in Germany. Something went wrong (perhaps Emil let the batch overheat), and a terrible explosion killed Emil and several others.

The fact that Nobel never wrote a line about how Emil's death affected him strongly suggests that it affected him all too deeply. What he did was to begin work on creating a stable form of nitroglycerin. Within a few years, he made his best-known invention. By soaking nitroglycerin into a porous packing material, he created a new substance that he called *dynamite*. Dynamite remains stable until it is triggered with a blasting cap.

In 1875, Nobel went a step further and created a jelly-like suspension of nitrocellulose and nitroglycerin, which he called *blasting gelatine*. The powder that he offered to the War Office had been derived from his blasting gelatine.

Nobel's inventions came on the eve of the great construction projects that followed Western colonial expansion. As we blasted railroads, tunnels, and canals across America, India, and Panama, Nobel grew wealthy.

But he did not grow happy. His childhood had been messy, he remained strongly attached to his mother, and he was subject to depression. He never married, and his sexual preference remains a subject of debate. Still, he had several intense, if distant, relationships with women.

One of those was his friendship with the Countess Berta Kinsky von Suttner, whom he met in 1876. When he was 43, Nobel placed an advertisement in a Vienna paper: It said, "A very rich, cultured, aged gentleman [seeks] an equally mature lady knowing foreign languages for the performance of the duties of a secretary and housekeeper."[7] (That business about being aged at 43 says a lot about Nobel.)

Berta Kinsky, a 33-year-old countess who had fallen on hard times, answered the ad. She had been in the service of the aristocratic von Suttner family until a romance had developed with young Arthur von Suttner. When Arthur's parents disapproved of her poverty and asked her to leave, she went to work for Nobel.

Nobel was smitten by the beautiful and intelligent Kinsky, and he soon professed his love for her. She told him, "Thank you," but she was engaged to Arthur. Soon after that, Berta and Arthur eloped. Nobel saw her only two more times, but they corresponded as long as he lived. As the Baroness von Suttner, Berta took up the cause of peace.

In 1889 she published a book entitled *Lay Down Your Arms*. The only other nineteenth-century book to become more popular was Harriet Beecher Stowe's *Uncle Tom's Cabin*. Berta von Suttner had touched a nerve and she had also touched Nobel. He wrote to her, "How long did it take you to compose this marvel? You will have to answer that question when next I have the honor and the pleasure to press your hand in mine— the hand of an Amazon who so valiantly wages war on war. . . ."[8]

The year before his death, Nobel created the foundation that would give the various Nobel prizes, and Berta von Suttner has been widely credited with stimulating the inclusion of a Peace Prize. Actually, the chemistry between them had worked somewhat differently. She had tried to draw Nobel into her various efforts to create world peace, but he had kept his distance—always insisting that his explosives advanced the cause of peace faster than any of her congresses. The day two armies could destroy each other in one second, he told people, all civilized nations would recoil from war in horror and disband their armies.

Well, dynamite did not accomplish that, nor have nuclear bombs. Still, peace appears to have been as constantly on Nobel's mind as it was on Berta von Suttner's. In the end, he did what he could to stem the damage that others had done with his inventions.

When he set up the Nobel Prizes, it was clear that they would include prizes in physics, chemistry, medicine, and literature. The Peace Prize was more problematic. Alfred's and Berta's letters argued over its form. Offering it for efforts at disarmament, he felt, would be futile; it should instead reward efforts at arbitration and ending the prejudices that caused war.

After Nobel's death in 1896, his will was hotly debated. Many people thought the Peace Prize was wasted money—foolish idealism. But it stood up in court, and the prize has been given ever since 1901. The first Physics Prize went to Wilhelm Conrad Roentgen for his discovery of X-rays, and the first Peace Prize to Henry Dunant for founding the International Red Cross. Four years later, Berta Kinsky von Suttner won the Peace Prize.

Maxim and Nobel represent, for me, two essential chemical components of *Modern*. Maxim was the savage boy–craftsman-inventor. Nobel, on the other hand, knew that his "historical neck," like that of Henry Adams, was being broken by "the sudden irruption of forces totally new." What neither lived to see was war in a form they both had done so much to create, and taking place on a scale that neither of them had conceived. Nobel would have been astonished when, far from recoiling in horror and disbanding their armies, nations would return again and again, with better powder and faster-firing guns, to create even larger wars.

The harbingers of modern war were there, of course. The use of flying machines, while it was on people's minds, remained abstract before 1914. More immediate was the combination of rifled and automatic weaponry with earthen fortifications.

Ask yourself what a fort should look like. The great American forts of the nineteenth century took a form that most of us have forgotten. Since the towers and loopholed overhangs on the old medieval castles could not stand up to cannon fire, forts had been redesigned by 1800 to present a simpler and lower profile—a long thin line of wall.[9]

Most nineteenth-century forts were square or pentagonal, with bastions on each corner. A bastion was a spade-shaped widening of corners that let defenders fire parallel to the walls. Bastions gave forts the shape of great stone snowflakes. The Confederate Army had built two such forts out on the far tip of the Mississippi, practically in the Gulf of Mexico.

They were Fort St. Philip and Fort Jackson. Each was a huge, low-lying building, surrounded by water. However, neither seriously affected the campaign in the West. In 1862, the Union fleet of gunboats subjected them to six days of pounding and then sailed on past them.[10] The Union took New Orleans, and those great forts did no more than slow the enemy down.

Heavy fortifications had already lost their role in war, but the military did not fully digest that fact. Two other actions that occurred just after the battle of Forts St. Philip and Jackson would display not only what was happening to large forts, but also the way in which forts were destined to

PRESENT ASPECT OF FORT JACKSON, FROM SUMMIT OF THE LEVEE LOOKING SOUTH FROM THE RIVER.

Fort Jackson on the Mississippi, as sketched in profile ca. 1885, from the April 1885 *Century Magazine*.

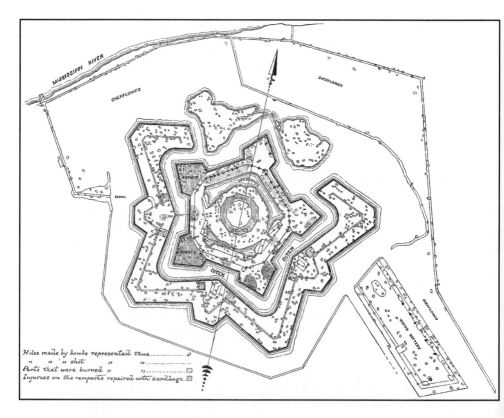

Plan view of Fort Jackson after the six-day Union bombardment, showing the extent of the damage, from the April 1885 *Century Magazine*.

be replaced. One was the fighting over Fort Sumter in North Carolina, which would last throughout the entire Civil War. The other was the Battle of Duppel in northern Europe.

Fort Sumter looms so large in American history because that is where the Civil War began. Located on the tip of an island in Charleston Bay, it is also five-sided, with the ocean coming right up to four of its walls. Only the fifth wall needed the protection of bastions. The first action of the war was the bombardment in which Confederate troops poured 40,000 artillery rounds into Fort Sumter. They killed no one, but, as the Fort began to crack, the Union forces surrendered.

The story of Fort Sumter was not over, however, for now the Union Army rolled out its new long-range rifled cannons. Those guns could sit outside the range of the old smoothbore Confederate artillery and pound Southern fortifications into gravel. In 1863, the North set up on Morris Island in Charleston Bay, beyond the range of Fort Sumter's guns. During the remainder of the war, Northern troops hammered the fort into rubble.

But the Confederates simply shoveled dirt and sand into the holes and burrowed in under it. In a few days the combination of Federal guns and Confederate ingenuity had turned the fort into an impenetrable redoubt. Period photographs show what appears to be total destruction. Yet Fort Sumter was now unassailable, and the North never did take it back from the South.

The Battle of Duppel was much the same story, but this time it was no accident. In 1864, Danes under attack by Prussian forces made their stand at the village of Duppel. There, six thousand ill-equipped Danes built a long, fortified trench and held off 18,000 well-equipped Prussians. Instead of crumbling under Prussian artillery fire, the Danes' protective dirt just flew into the air and fell back down. The Prussians won, but it took two months, and it cost far too many lives.

Poverty of means thus did for the Danes what accident had done for the American South. We are thus left with a remarkable example of how technology informs us. Masonry was no protection against the new rifled cannons, but pulverized walls were. By the end of the Civil War, the South had reinvented defensive fortifications. The old masonry forts lingered another half century, but they never again played any important role in war.

When Barbara Tuchman wrote her famous book about World War I, she titled it *The Guns of August.*[11] She did so because, 50 years after Sumter and Duppel, the military had still not adequately read the lessons the battles had to teach. As a consequence, World War I took on a completely unanticipated character during the first few days of August 1914.

The German attack plan, from a nineteenth-century military strategy book, promised victory in a few weeks. Instead, war soon mired into four years of human attrition along a double line of impregnable dirt trenches running from Nancy west almost to Paris, and then north to Ostend. By the time the carnage was over, 8.5 million soldiers had died, most of them attempting to dislodge the other side along this 350-mile line.

Both sides tried to fight the war much as Napoleon would have fought it, despite a half century of warnings that the new technologies of slaughter would create a stalemate. The storm warnings had kept right on coming after our Civil War. Open-field formations were cut to bits in the Prussian-Austrian conflict in 1866, and again when the Russians fought the Turks in 1877. Napoleonic muskets had evolved into such rapid-firing rifles as the British Enfield, and Maxim's gun had matured.

These weapons laid down a field of fire that made open-field attacks impossible. Once entrenched, armies were practically immovable. Five thousand soldiers were killed during a slack week along that double row of trenches; a half million could die in a pitched assault. Yet the line hardly moved. The Allies finally managed to get around the stalemate with their new armored tanks, but only after they had won this war of attrition.

War itself was a collision between the nineteenth century and *Modern*. But *Modern*, in its purest form, quite literally rose above the war to take part in it. *Aerial* warfare was almost without precedent in 1914. In the early 1860s the Union Army had used balloons for observation, but not for combat. In the Franco-Prussian war, the Parisians had used balloons to communicate with the outside world during April and May of 1871, while they were cut off by the German siege.[12]

A new era of aerial warfare was widely anticipated in the years leading up to 1914, but it was also speculative. Discussions of aerial warfare took the form of a new science-fiction genre. One is left wondering how seriously people took what they read. The airplane certainly was immature enough to be underestimated, even though it was evolving very rapidly. The dirigible, on the other hand, was already very impressive. If any of the machines now in the air appeared ready to play a role in war, they were the great airships.

For years before World War I, magazines dramatized aerial warfare as writers imagined it would take place. The very theatricality of those prophecies and images probably caused people to view them as unreal. They captured the sense of what was to come, but the details were inaccurate, and they ultimately underestimated the real potential for destruction by aerial bombardment.[13]

The German military, however, *did* take aerial warfare seriously. The Germans contracted with Count von Zeppelin to build Zeppelin dirigibles for military service, and he went into production in 1906. Then a disorienting thing happened:

Germany's Krupp weapons factory had long been a central pillar of German nationalism, fiercely loyal to whatever regime was in power.

"War in the Air" as envisioned in the 1909 *McLure's Magazine.*

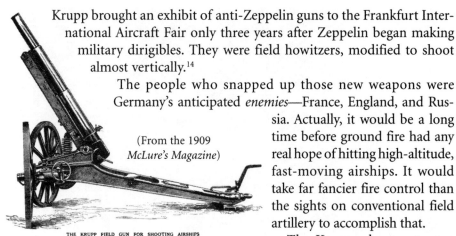

Krupp brought an exhibit of anti-Zeppelin guns to the Frankfurt International Aircraft Fair only three years after Zeppelin began making military dirigibles. They were field howitzers, modified to shoot almost vertically.[14]

(From the 1909 *McLure's Magazine*)

THE KRUPP FIELD GUN FOR SHOOTING AIRSHIPS

The people who snapped up those new weapons were Germany's anticipated *enemies*—France, England, and Russia. Actually, it would be a long time before ground fire had any real hope of hitting high-altitude, fast-moving airships. It would take far fancier fire control than the sights on conventional field artillery to accomplish that.

The Krupps, however, were also looking after Germany. The same year they presented their anti-Zeppelin gun, they also unveiled a gigantic, 94-ton mortar with a foot-and-a-half muzzle. It was nicknamed *Fat Bertha* or *Big Bertha* (*Dicke Bertha,* in German). That was in deference to the beautiful Bertha Krupp who had inherited the Krupp works in 1902 at the age of only 16.

It was unthinkable for the titular head of what was effectively the industrial arm of the German empire to be a woman, so Kaiser Wilhelm himself helped to broker a quick marriage between Bertha and one Gustav von Bohlen und Halbach. At the wedding, Wilhelm announced that Gustav would henceforth use the surname of Krupp.

Twelve years later, war: Big Berthas tore Belgium apart and then reduced much of the French landscape to mud. Bertha Krupp resigned herself to the name Big Bertha, for, after all, that was a part of her job. She was still alive after the next war, when she watched her son Alfred being tried at Nüremberg for his brutal use of slave labor. He got off, because America intervened. America now needed him to rearm Germany against the threat of Russia.

As railroad cars moved the Big Berthas into position, the Germans also put their Zeppelin dirigibles to work as bombers. Large as ocean liners and four times as fast, they initially promised to be terrifying weapons. Two of them reached Yarmouth in January 1915. They damaged the town square, killed four people, and wrecked some small buildings. The next month Germany tried again and lost two Zeppelins in a storm over Denmark.[15]

Not until September 8, 1915, did Germany have its new bomber tactics under control. That night, four dirigibles rose up over the North Sea and made for London. Their prime target was the Bank of England. The raid did great damage to civilian property and left 109 casualties in its wake.

Official English reports made light of the damage, but a German spy got hold of a secret report. It said the airships flew too high for English planes to reach or for searchlights to hold them. It also admitted that the damage had been far greater than the British had told the public. Germany was delighted and went on to expand its airship tactics.

For example, Zeppelins began using an observer capsule that they could lower thousands of feet while the airship lurked in cloud cover. That was a nasty assignment; one observer was dashed against a cliff. Another was jettisoned when the dirigible suffered a gas leak and a stuck winch. This time, they managed to cut the cable just after the sinking airship was low enough that the capsule struck earth. The observer tumbled free and spent the next two months dodging the English, living in barns, and stealing food. He actually breathed a sigh of relief when he was finally caught.

As Germany, fueled by the illusion of success, poured its efforts into dirigible warfare, better fighter airplanes began reaching dirigible altitudes. Defense technologies improved faster than attack technologies. The cost finally became too high, and Germany had to give up on Zeppelin bombers.

Yet dirigibles foreshadowed the contradictions of bombing that remain to this very day. They did more civilian damage than most history books tell; but, like the Battle of Britain in 1941 or the bombing of Hanoi in the 1970s, they awakened resolve in the enemy below. In the end, the more effective aerial warfare would be shaped around airplane tactics above the trenches.[16]

Generals on both sides were dubious of airplanes at first. The old technology of observation balloons was familiar, and they used it. Both sides had balloons, and few people saw that the flimsy airplanes of 1914 would improve upon them by providing moveable aerial platforms. When French scout planes reported a German buildup around Verdun, their generals scoffed at the idea that airplanes could reveal what normal intelligence could not.

France almost lost the war at that point. The military was still thinking only in terms of Zeppelin bombing, even though, for low-level tactical work, dirigibles not only became vulnerable to the most rudimentary fighter airplanes; they became vulnerable to ground-fire as well.

The first task of airplanes had likewise been tactical bombing, although that story was not widely told. Their devastation of troops in the trenches made poor press back home. The first airman to win a Victoria Cross was a flier named Rhodes-Moorhouse. He dropped a one-hundred-pound bomb on a railway signal box and slowed the movement of German troops into the Ypres sector. He brought his airplane home, but he also carried a fatal German rifle bullet in his stomach. Airplane bombing played a significant role at sea as well—damaging, and even sinking, some large ships.

Airplanes thus established their roles, first as bombers, but then also as scouts. As they did, their pilots looked for ways to get rid of enemy airplanes. They began dropping bags of bricks and metal darts on each other. They tried dangling chains into enemy propellers. They began carrying pistols. The French were the first to mount machine guns on airplanes. At first, the weight of a Lewis gun sorely taxed an early airplane's load-carrying capabilities. Airplane designs had to catch up with this new purpose.

Much adjustment began taking place immediately after August 1914. It took about a year for the maneuverable airplane with forward-firing guns to become an icon of the war. Its obvious purposes would be to scout enemy positions and shoot down observation balloons.

But the aim of eliminating one another also emerged as a primary imperative. So too emerged the new image of the pilot as the knight-errant of *Modern*. The real Red Baron, Manfred von Richthofen, began awarding himself a silver cup each time he downed an airplane. He collected almost a hundred before he, too, was killed.

The story of World War I fighters says a lot about the rhythm of technological change. It had a kind of speed and intensity—an alternation between complacency and desperation—that we see in the computer business today. Early in the war, the very young Dutch airplane maker Anthony Fokker emerged from a gaggle of competing makers of German airplanes with his single-winged Eindecker.[17] The Eindecker was a very robust airplane for those times. It flew just over 80 miles an hour, and it could do some pretty fancy aerobatics.

Then, in 1915, Fokker invented an interrupter mechanism that stopped a machine gun from firing each time the propeller blade passed in front of it. That way the gun could be mounted directly in front of the pilot and could fire through the moving propeller with devastating accuracy.

France answered with a biplane, the Nieuport 10. It performed well, but its machine gun, mounted on the upper wing, still had to fire over the propeller. It was hard to aim, and the pilot had to stand up in his cockpit when it jammed and needed fixing. By early 1915, many early Nieuports had been equipped with synchronized machine guns. The Nieuport 17 appeared a year later, and, for a while, it seemed to have leveled the playing field.[18] But Fokker was not done yet. Now he built an airplane with not one wing, but *three*.

In fact, he was copying an earlier Sopwith triplane. The reason for extra wings was not to add lift but to gain the structural solidity of wings

Fokker Eindecker with a forward-firing machine gun. (Family photo provided by
Karl-Heinz von Beaulieu-Marconnay)

braced against one another, along with greater maneuverability. Fokker's
triplane had much shorter wings than either the old Eindecker or even
the early Nieuports. By 1917 it was doing such damage to the Allies that
the French tried to build a three-winged version of the Nieuport 17. But
it took the arrival of another biplane, the highly maneuverable Sopwith
Camel, to level the playing field once again.

Three formidable entrants appeared in the last months of the war:
the Fokker D-VII, the Nieuport 28, and the SPAD. Now the formula was
worked out, and all three were similar. They were 120-mile-an-hour bi-
planes, with wingspans just under 30 feet. It had taken four years of
tinkering and thousands of deaths to settle on that design. The path was
littered with far more added and removed parts than any airplane buff
can list. Every field mechanic was part airplane designer.

But the early design that came closest to the final form may well have
been that neat Nieuport 17. My father told me that it was the apple of
his eye. He flew all kinds of planes in the last months of World War I—
Curtiss Jennies, Sopwith Camels, SPADs.

I count myself lucky that he arrived too late to get into combat, be-
cause by then those pretty Nieuports had been outclassed, first by the
Red Baron's Fokker triplane, and finally by the superb Fokker D-VII.
However, the seminal success of the Nieuport 17 stemmed from the fact

Postcard from Lt. J. H. Lienhard to his mother, 1918. The airplane with the British markings is an imperfect artist's conception of what was almost certainly intended to be a Nieuport 17.

that it had already captured the essential form of the last World War I airplanes—two equal wings and a clean, uncluttered shape.

And so, on October 14, 1918, my father reached Issoudun, in Central France, where the Allies had set up a huge flying school. He trained there until a few days after the armistice. Then he was sent on to the 185th Night Pursuit Group, about 20 miles from Verdun.

On November 9, just two days *before* the armistice, a strange thing had happened only a few miles away from that airfield. A tiny American airstrip was located just east of Verdun. All one could see were some sheds. That day the airplanes were in the sheds, and three pilots were playing cards. They were Alexander H. McClanahan, Edward P. Curtis (who later became a vice president of the Eastman Kodak Company), and Sumner Sewall (who later became governor of Maine).

Suddenly, they heard a sputtering engine. A Fokker D-VII dropped out of the low gray clouds and landed. The American pilots ran out with pistols drawn and captured the German pilot. He was Heinz von Beaulieu-Marconnay, a Huguenot aristocrat. They invited him in for a collegial shot of cognac before they sent him off to be processed as a prisoner of war. Afterward, the three airmen flew von Beaulieu-Marconnay's Fokker D-VII and then delivered it over to Issoudun to be shown off.

The airplane was eventually boxed up and sent to America, where it wound up in the Smithsonian Institution. In the 1970s the museum restored it to its 1918 condition, which included a mysterious "U10" painted three feet tall on its side. I was studying in the Smithsonian during the summer of 1970 and was given the task of writing a museum label for that airplane.

I knew about the Fokker D-VII, but what on earth did U10 stand for? As I looked, the plot thickened. Two weeks before the armistice, the pilot's brother, a German ace named Oliver von Beaulieu-Marconnay, died of wounds after he was shot down. His plane had written on it the equally mysterious, "4D."

And why had Heinz landed? Was he broken by his brother's death—saving himself by surrendering? That was current Smithsonian think-

Heinz von Beaulieu-Marconnay's Fokker D-VII at the National Air
and Space Museum, with the once-mysterious U10 marking on the side.
(Photo by John Lienhard)

Lt. Heinz von Beaulieu-
Marconnay.
(Family photo provided by
Karl-Heinz von Beaulieu-
Marconnay)

ing. I went looking for answers to the questions created at that remote airstrip, on an overcast day late in the war.

I located S. Lawrence Levengood, the then-elderly American interrogator who had talked with Heinz soon after he landed. I was astonished to learn that he and Heinz had become close friends after the war. Nothing that Levengood had learned remotely suggested either depression or defection. Heinz was reconnoitering, had had engine trouble, and decided to put down on an airstrip that looked unoccupied. A letter from Levengood (sent to me on November 5, 1971) made that very clear, and it included these poignant and remarkable passages:

Heinz was always gracious, courteous to all about him, a devoted husband and lenient father and one of the best friends that I have ever had. I mourn his tragic death as a Russian prisoner in Siberia after World War II, as I grieve for the death of his lovely wife Maria last January.

[During our interviews, we talked] of visits that I would make in Germany, which I am glad to say did take place most happily. I became Onkel Larry to the three children and, I think I told you, godfather to Karl-Heinz, the oldest. I was also at the christening of Christa in 1929.

Levengood alludes here to Heinz's later history. He became an important figure, first in Lufthansa, and then in the Luftwaffe after the military took over Lufthansa. He was captured again during World War II, and he died (reportedly of pneumonia) in a Russian prison camp.

I also found Heinz's children. His daughter was living on Long Island. His son did drafting and design for the Messerschmidt Company. Neither could say what Heinz's U10 or Oliver's 4D stood for. Finally the daughter wrote to her aunt, who said that both her brothers had joined the cavalry when war began. Oliver went to the Fourth Dragoons, hence the 4D, and Heinz had ridden with the 10th Uhlans regiment.

When the brothers had the chance to fly, they had seized it. They turned from nineteenth-century war to modern war, and they took their cavalry insignias with them. They left the mud on the ground below and sought adventure. Oliver had become a major German ace, with 25 victories. Heinz also scored a couple of victories, but not enough to qualify him as an ace.

So my quest had taken me into a twilight zone, a last vestige of nineteenth-century views of war combined with a radical leap into *Modern*. Here we find the thin tissue of a chivalrous war, a gentleman's war, a war that never was, woven into war as it would become. It was a strange world that comes back to us again in a passage from my father's diary.

His mind had been focused on flying and the shiver of anticipated aerial combat. Now, four months after the armistice, he had a chance to fly as a passenger in one of the late-arriving DH-4s (see chapter 11). He is suddenly given his first look at what had been happening down on the hard earth of no-man's-land. He writes:

Jan 10 1919

Had the ride of my young life yesterday. Went up in the DH4 (Liberty) with Romaine as pilot and we took a "Cooks tour of the trenches". That is, we went up the line to Fluery, over to St. Miheil, up to Metz and after an extended jazz over Alsace Lorraine, back down over hill 360 and Verdun. The trip took 1:45. Most of the country we went over was rolling countryside with the fields of gray, brown and vivid green, brown forests and twisting rivers but nearer home we had to pass over belts of country which had been ruined.

Around hill 360 and Verdun the country is torn by shells in a way that is scarcely believable. It looks just like a mud bank would look if one were to throw a handful of shot or pebbles into it—all full of round dots or spatters—shell holes so close together that they for the greater part overlap.

There was something raw and wild about 360—it struck me as being more impressive than the fortresses about Verdun. These are huge cement inserts set into the hills above the town—I only remember seeing two of them. The villages in this neck of the woods are simply stone piles with here and there a piece of wall standing. Metz is a pretty place and stands out clean and bright on the banks of the Moselle.

I took some pictures from rear seat but dont know how they will come out. The Liberty performs well—has all the speed and power on earth but there is an enormous wind pressure and you get the full benefit of it standing up in the rear seat.

When my father looked down from that DH-4 there was little of chivalry—no more camaraderie of warriors sharing shots of cognac—in the landscape below. This war, like all wars, had really been about killing

Shell-Holes by Orville Houghton Peets, from the November 1918 *Century Magazine*.[19]

and destruction. But now it had all been taking place on a new and unprecedented scale. My search for the Fokker D-VII story had put me into contact with many very decent people who had, ultimately, been trying to make the war into something quite other than what it really had been.

A strange disjunction was now taking place as *Modern* found the sky above. In the sky, war could be antiseptic. It was all a matter of which way one looked. Above was blue sky, below was mud. The dreams are to be found above, and mortality below. *Modern* now meant that the machinery of war would increasingly be found in the skies, even if death remained earthbound.

The German Army had been first with military airplanes, and, in 1929, it became first to take on military rocketry as well. The army ordered a study of rockets in 1929 and began building rockets the next year. That sounds like great foresight, and, I suppose, it was. But it had nothing to do with the Nazis. Hitler did not take over until four years later.[20]

By 1929 the American, Goddard, and the Russian scientists had worked on rockets for the sheer fascination of the potential accomplishment. The German military had not tumbled to any of that. Instead, we have to go back to 1923, when Hermann Oberth published a book on rocketry. Oberth laid out the principles of modern space flight. He said that rockets could escape Earth's atmosphere; they could even escape gravity; they could carry people; and they could turn a profit.

Oberth was not talking about war; he was interested in *space travel.* The book was to have been his doctoral dissertation, but the faculty at Heidelberg rejected it. Rockets would not work in the vacuum of space, one professor sneered, because they have nothing to push against.

Then a science-fiction writer named Max Valier got his hands on Oberth's book and became Oberth's champion. Writings on space flight began flowing from his pen. Valier arrested public attention. For a while, Valier and Oberth collaborated. But Valier did not really understand rocketry. He tried to improve on Oberth's hard facts, and Oberth had to end their collaboration.

Facsimile of Goddard's rocket at NASA JSC.
(Photo by John Lienhard)

The spark had, however, landed in the tinder of the modern imagination. By 1928, Valier's enthusiasm had reached the heir of the Opel automobile fortune. The fellow built a series of rocket-powered cars, and then he flew a rocket-powered glider. Meanwhile, Fritz Lang stopped work on the movie *Metropolis* to make a film about flying to the moon.

In the midst of all this, a German officer read Oberth's book. Only then did that officer start the machinery of military rocketry. By 1930, we find a photo of the military rocket group. Oberth stands in front of a small rocket. On the right we see an 18-year-old lad holding a fitting with his clear, alert face gazing, transfixed, not at Oberth, but at the rocket. The young engineer is Werner von Braun.

Fourteen years later, V-2 rockets began falling on English civilians. They were not the fruit of military vision at all. They were born in Oberth's science and Valier's fiction. They were born in the face of young von Braun, dreaming about flying to the moon.

And so invention does not rise to the necessities of war at all. Rather, the necessities of war, now and then, provide creative minds with the outlet they seek. Just as Krupp created the anti-aircraft gun and then sold it to Germany's enemies, von Braun looked at the sky overhead, not at the ground below, and he lived to see astronauts on the moon. I used to think I understood the morality in all this better than I do today.

Nowhere are those issues more devastatingly evident than they are in another German pilot, Hanna Reitsch. Reitsch was born in Germany in 1912. She wanted to be a flying missionary doctor, but after the Versailles Treaty had clipped Germany's wings, she became an excellent glider pilot. She set records, she worked as a movie stand-in flyer, and she went on an expedition to study weather in South America. Hitler made her an honorary flight captain, the first woman to receive that award.

In 1937, the reformed Luftwaffe hired her as a civilian test pilot. She accepted with near reverence, calling German warplanes "Guardians of the portals of peace." Historian Judy Lomax tells how Reitsch's values were instilled by a mother who wrote her daily, warning against the sin of pride and praying for her safety.[21]

Reitsch rode the forefront of German technology. She tested the gliders, which silently deposited German troops right on top of France's first line of resistance, the Maginot Line. She tested Germany's first helicopter. She worked fervently and methodically in a cause she accepted without question.

In 1941, Hitler awarded her the Iron Cross, second class, for the almost fatally dangerous work she did in developing means for cutting the cables dangled by British barrage balloons. The most dangerous machine that Reitsch tested was the Messerschmitt 163, Germany's experimental rocket-powered interceptor. In a minute and a half after takeoff it climbed at a

65-degree angle to 30,000 feet. It traveled nearly 600 MPH, the fastest any human had ever gone.

On her fifth flight, the takeoff dolly jammed. She crash-landed, split her face open, and still had the presence of mind to write out what had happened before she passed out. She spent the next four months hovering between life and death; and, this time, Hitler gave her the Iron Cross, *first* class. When she recovered, she was horrified to learn that the Me-163 was in full production. She knew that was crazy wishful thinking, that the plane was as useless as it was dangerous.

If that did not shake her belief in the Nazis, rumors that they were exterminating Jews might have. But when she confronted Heinrich Himmler, he made her believe he was as outraged as she was that the Allies would spread such propaganda. And so Reitsch remained a believer. When she learned that Germany was thinking about a suicide version of the V-1 rocket—a Kamikaze bomb to be guided by a human being—she asked to test the prototype. She was disappointed when she found it was an empty threat.

After the war she was doggedly unrepentant. She wore her Iron Crosses proudly and wrote somewhat defensive memoirs. Was she a Nazi to the end, or just a proud woman? We don't know. She continued to fly and was generous in helping other women pilots from other countries. At the age of 65, the year before she died, she set a new women's distance record in a glider.

Before it was done, *Modern* would show us yet worse weapons than any Hanna Reitsch had tested. Those weapons, by and large, would continue to seek the sky above. Writer Alexander Roca describes one invention that, for me, puts *Modern* and war together in microcosm.[22] It began during World War II, when chemical warfare developers went to inventor Thomas Shelton and asked him to invent a way to throw poison gas from a distance. Because the gas diffuses so quickly, squirting it from a nozzle will injure whoever is holding the nozzle.

Shelton, well-known for his knowledge of vortex shedding in airplane design, sat down and invented a large, bell-shaped chamber. When he exploded a shotgun shell in one end, it burped a gigantic smoke ring out the other. A smoke ring is a vortex of spinning gas that holds together while it propels itself forward.

This particular vortex was 18 inches across, and it traveled 150 feet. According to Roca, the gadget made "an eerie howling sound." From there, Shelton went on to interest the Air Corps in a huge version that would supposedly fire vortices large enough to knock enemy airplanes out of the sky. In the end, the war effort went, as it usually does, in safer and more conventional directions, leaving us to wonder whether Shelton's big vortex guns might have had some grain of plausibility after all.

Since his weapon had come to nothing, Shelton turned it into a toy gun after the war. With it, a child could use an air vortex to knock over targets. He called it the "Flash Gordon Air Ray Gun." *Popular Mechanics* named it "Toy of the Year" in 1949, and the Lion Match Company invented a pellet that generated smoke for the vortices.

Shelton's Air Ray Gun was the ultimate exemplar of *Modern*. It had all the earmarks; it smacked of the boy mechanic, the boy scientist, and the better mousetrap. It was streamlined, and it carried the proper patina of mystery. War was behind it, but it was done with war and was now involved in play. It was something Hanna Reitsch might have enjoyed playing with when she was a little girl. The box says:

> The most amazing toy of the modern age
> Absolutely safe—harmless
> Shoots a whizzing jet of harmless air

The Flash Gordon Air Ray Gun is somewhere out there with automobile tailfins as an image of *Modern* on its last legs. *Modern* and war were, without a doubt, a marriage made in hell, and by 1950, the public hoped to leave war behind and to embrace all that *Modern* had been promising before war began.

Alas, the opposite would be the case, for now we had the bomb. We had the ICBM. During the 1950s it was *Modern* that we would leave behind, and war that would remain with us. What happened during the 1950s? Where did *Modern* go? Let us turn that particularly intractable question about in the light, and see what we can do with it.

15

A Funeral in the Fifties

*M*odern breathed her last during the 1950s, and we witnessed her passing without any real sense that she was ailing. Such a peculiar alchemy of change and contradiction enveloped us all—we were done with World War II and back to life where we had left off. How could anyone have realized that we were now somehow switched over to a completely different train, and that we were riding down an entirely new track?

Fifty years later, I still recall trying to cast my world into familiar terms and only gradually realizing that 1939 had vanished. I tried to see life as the logical technological extrapolation of everything familiar, and it kept turning into something else. The old technologies were still there, still indexing ahead, but somehow the subject of the conversation had been changed, and they were no longer at its center.

For me, those years remain very raw. They were a piece of my life that I cannot, even now, view with any objective detachment. I shall, therefore, drop all pretense of analysis, and present myself as a primary source—a witness, a cork on the water. Just as I offer my great-grandfather as an "everyman" of 1846 (chapter 1), I myself play that role here. I make my deposition on the end of *Modern*, knowing full well that, if you were there as well, you might want to stress other events or stress my events differently.

I begin near the end, in the years from 1956 to 1961, when I was a graduate student at the University of California at Berkeley. Part of my dissertation was connected to a study funded by the old Atomic Energy Commission, which had asked us to find out whether or not the vapor generated by boiling during a transient energy excursion in the core of a power reactor would shut down nuclear fission and protect the reactor.

My mechanical engineering background suited me to work with the thermofluid action of boiling, but I needed to learn more about the nuclear processes involved. So, in 1960, when my friend Balraj Sehgal taught the introductory nuclear engineering course, I sat in.[1] One day Balraj made an odd remark to the class. On some level, more intuitive than objective, he captured what had happened in the late 1950s. But before I quote the remark, I should explain the context:

Western science and technology arose upon the Neoplatonist philosophy of the early Christian Church. However, since Plato had provided no body of science, the Church incorporated Aristotle's natural philosophy into that system of knowledge. Aristotle had, in turn, picked up the existing idea that all matter is composed of the four essences identified with earth, air, fire, and water. He explained that heat and cold, dampness and dryness, act upon those essences to give matter its form.

Medieval alchemists rediscovered Aristotle and began doing some very fruitful work on the transmutation of matter into different forms. They set the foundations of process chemistry, and, because they had accomplished so much, the Aristotelian essences were slow to give way to atomism. That shift did not begin until the seventeenth century, and I mention in chapter 3 how the full acceptance of an atomic description of matter was resisted until the late nineteenth century.

Yet, no sooner was atomism accepted, than it began developing fuzzy edges. By the late 1920s, physicists were already describing small particles as quantum "waves of probability." Pure bowling-ball atomism had set the stage for *Modern,* and then, almost immediately, it began fragmenting.

And so Balraj, talking more to himself than to the class, remarked that *our seemingly solid atomic world was actually morphing back toward Aristotle's continuous essences.*

His remark is important in retrospect, precisely because the public of the 1950s had begun looking upon nuclear engineers as latter-day alchemists, dealing in the black arts. Nuclear engineers quite literally *do* transmute matter into energy. Had I been more alert when Balraj said that atomism was blurring back into Aristotelian essence, I might have realized that a similar message was sounding on the faint but restless drums out in the jungle of public opinion.

Balraj, my wife, and I hung out together, and one of our haunts was a theater down in Oakland, the New Peerlex. The New Peerlex showed every B-minus science-fiction movie as it came out: *Attack of the 50-Foot Woman, Plan 9 from Outer Space, This Island Earth* . . . It became our decompression chamber.

We would quit working at around 8:30 on Saturday evenings and go to watch a late-night triple feature. One of those movies in particular stayed with me. It was the Ed Wood classic, *Bride of the Monster*—a masterfully dreadful movie by any standard, and one of Bela Lugosi's last appearances.

Lugosi, far beyond his prime, delivered an unforgettable line that perfectly captured what was going on at the time. Toward the end of the movie, mad scientist Lugosi turned upon the camera with terrible intensity, and said, "Home. I have no home. Hunted . . . despised . . . living like an animal—the jungle is my home! But I will show the world that I can be its master. I shall perfect my own race of people—a race of atomic supermen that will conquer the world!"

We chuckled and went back to work, not at all understanding how powerfully Lugosi (and all the other mad scientists) were speaking to the public. Balraj solved a form of the Boltzmann equation, and I built my Faltung integral for the fraction of vapor present as water boiled in the reactor core.[2] Despite Balraj's remark in class, he and I behaved as though we were, in fact, still dealing with clean atoms. We kept trying to live in a world whose edges were well-defined.

But the public has a nose. The public knew that the world really *is* made of mysterious essences after all, and that those essences really *could* fall into the hands of Bela Lugosi. The transmutability of matter and energy really *does* represent a return to Aristotelian alchemy. Alchemy really *had* returned, along with quantum mechanics and relativity theory.

Long before we had the bomb (or Lugosi's race of atomic supermen) to worry about, scientists knew that we would one day transmute matter into energy. As early as 1905, Einstein had shown us that we would be able to extract the pure alchemical principle of fire from base earth. *Modern* was just beginning when his formula, $e = mc^2$, expressed the fact that mass, m, and energy, e, may be obtained from one another.

The pivotal event in any history of atomic power came much later, in the definitive work of Otto Hahn and Lisa Meitner, which culminated in 1938 and 1939, respectively.[3] Meitner had received her doctorate in Vienna and, in 1908, went to work for Max Planck in Berlin. She and her colleague, Hahn, remained close for over 60 years. Women were not allowed to work in laboratories in those days, so Hahn and Meitner created their own laboratory in a carpenter's shop.

They worked on nuclear fission until World War I. Then Meitner joined the Austrian army as an X-ray technician, but she kept on working with Hahn whenever they both could get away on leave. By 1918 they had revealed a previously unknown element, which they called *protactinium*. You might not recognize it, but it is there on your periodic table.

By the end of World War I, Germany had lost much of a generation of males, and opportunities briefly improved for women. Meitner was made head of the physics department at the Kaiser Wilhelm Institute in Berlin, where she and Hahn could work on the effects of α- and β-radiation. Sixteen years later, they were bombarding heavy elements with fast neutrons. Meitner finally saw what an enormous energy release occurred when her bombardments caused uranium to fission into barium.

But she was both Jewish and a committed pacifist. She had to find a twisted path out of Germany. First, she fled to Holland on an invalid passport, then to Niels Bohr's home in Copenhagen. She finally got across the North Sea to Sweden, just ahead of Nazi patrol boats. There she published a clear explanation of nuclear-fusion energy in 1939. Her paper expressed hope for a "promised land of atomic energy." Her aims had nothing to do with bombs. However, as World War II began, her paper launched furious bomb-making efforts among the warring nations.

The Allies got through to her and asked her to join the Manhattan Project, but she would have no part of it. Six years later she was appalled to see the devastation her work had wrought in Japan. Years afterward, Meitner became the first woman to receive a share of the Fermi Award for her physics and, implicitly, for her contributions to the bomb she had never wanted to see made. Now 88, she begged off, saying that she was not up to the trip. Glenn Seaborg went to London, got the prize, and brought it to her.

Meitner and Hahn had merely provided the experimental proof of something that the public had known about for a long time. Before I was born, and long before Meitner's paper, my father worked on the *Fargo Forum* in North Dakota. In November 1925, as a 32-year-old feature writer, he wrote this:

Dr. Gerald L. Wendt, dean of the Pennsylvania State College school of chemistry and physics, has made the first small beginnings toward unlocking the atomic power which is in all matter.

This would mean that man would have unlimited energy at his command. "If I succeed in my endeavor, it will mean the revolution of civilization," says Dr. Wendt.

Mining of coal would cease, railroads would be electrified, cities would dissolve into great suburban areas and all heating, cooking and ironing would be done by electricity—probably the easiest way of applying the power.

My father makes Wendt sound a bit like Bela Lugosi there, and he goes on to say something that mirrored the public reaction:

By all means uncover this source of unlimited power, if possible.

But don't expect too much from it. After all, we've got more energy now than we have the wisdom to handle properly. Infinite energy or power would mean a little change, probably for the better. But a little more wisdom means infinite improvement.

The press had been speculating about atomic energy for years, even before World War I. By then, *Popular Mechanics* and *Collier's* had talked about atomic autos and airplanes. When World War II brought the lid of secrecy down on all that talk, I heard only one more thing about atomic power before Hiroshima, although at the time I had no idea it was connected.

A drunk called my father one night: "You guys at the paper wanna know what they're really doing out in Hanford, Washington?" "Sure," said my father. "Well," he slurred, "they're making front ends of horses for final assembly in Washington, D.C." Later, we would learn that they were, in fact, preparing uranium for use in the first atomic bombs.

Once the bomb was out in the open, and we saw what we had done, people frantically tried to turn attention back to civilian atomic power. Fine scientists became absurd predictors. Glenn Seaborg promised atomic airplanes. William Laurence promised sightseeing rockets to the moon and Mars, and Alvin Weinberg (first head of the AEC) agreed. Laurence also said, "[Splitting] the atom can lead to such priceless boons as the conquering of disease, the postponement of old age and the prolongation of life."[4]

It was a moment of childlike hope in the wake of war. But then J. Robert Oppenheimer, that knight of the woeful countenance who had headed the development of the bomb, pointed out terrible technical problems. Luis Alvarez warned that using atomic energy to fly a plane was "to do an easy thing the hard way."

As Balraj and I worked on problems of safe core cooling, it became clear soon enough that we would not see nuclear reactors in automobiles or airplanes for a long, long time. Instead we heard the voice of Bela Lugosi at the New Peerlex Theater promising that atomic power would bring terrible transmutations of the human species.

The movie *Them* promised 30-foot-long ants. We were back to alchemy: if we could transmute uranium, why couldn't we just as well transmute a frog into Godzilla? The public grew *very* afraid. Science and high technology had been so full of hope. Now, for a season, we were afraid of the bomb and afraid of transmutation gone rampant. We were afraid of the Russians.

If we had been celebrants of genius in the early twentieth century, genius now seemed to be poised to turn upon us. Benjamin Britten's *Hymn to St. Cecilia* became a very popular work after the war. It was based upon an older W. H. Auden poem that he rewrote for it. The text included two lines that fit the shifting view of the fifties like a glove:

> Impetuous child with the tremendous brain,
> O weep, child, weep, O weep away the stain, . . .

The bloom of *Modern* was off the rose. I left graduate school in 1961 and began the slow process of coming to grips with the fact that I had entered a postmodern era.

The storm warnings had been up ever since the end of the war, and they had all been about fear. Our fear of the bomb was all tied up with our fear of the Communists. I first encountered that fear ten years earlier, when I finished my bachelor's degree in 1951 and went to the Boeing Airplane Company in Seattle.

Later, I learned that, in 1947, the Washington state legislature had created a Washington State Committee on Un-American Activities, headed by state senator Albert F. Canwell. Canwell began by "rooting out communism" at the University of Washington. That fall of 1951, when I began working part time on my master's degree, I found that the committee had managed to fire three professors and put 30 or so more on probation for supposed communist leanings. Next, it had turned on the Seattle arts community and done comparable damage.

But I knew none of this when, in the spring, I looked for a place to live on my $280-per-month salary. I eventually found the All People's Student Center, just across the street from the University of Washington. The people there included every element of society that lay outside of acceptance. In addition to a dense mixing of races and nationalities, there were two ex-convicts, a couple of former card-carrying Trotskyite Communists, some homosexuals, and (most important) several people who had already been grievously wounded in varying ways by America's rising fear of the Red Menace. I was Candide wandering into a world that I thought I could observe from outside.

Fred Shorter, who ran the All People's Student Center, also ran an adjacent enterprise called the All People's Church. Fred was a former Congregationalist minister who had been defrocked during the late 1930s for what later became known as *premature antifascism*. (He had brought about his defrocking by allowing his youth group to paint anti-Hitler cartoons on their clubhouse wall.) By the time I arrived, Fred had become a leading civil rights activist.

I finished my degree three days before I was drafted into the U.S. Army in 1953. I figured, in my naiveté, that I was simply walking away from a very interesting episode in my life. But, of course, that was not to be. Canwell and a handful like him had gained the attention of the junior senator from Wisconsin, Joseph R. McCarthy, who, in 1950, was casting about for a way to make his mark upon the Senate.[5]

On February 9, 1950, McCarthy made national headlines with a speech given in Wheeling, West Virginia, in which he claimed to have a list of hundreds of communists in the State Department working under the direct control of Moscow. His red purge continued from that point until after he turned his guns upon the Signal Corps Laboratories at Fort Monmouth, New Jersey.

McCarthy's Fort Monmouth campaign began just as I started basic training at Camp Picket, Virginia. By November 1953, he had managed to suspend 33 civilian employees. I was assigned to work in the Signal Corps Laboratories at Fort Monmouth in April 1954, and there I found essentially the same climate of fear that had greeted me when I arrived in Seattle three years earlier.

On leave in Seattle that summer, I ran into a friend who had once belonged to the Trotskyite Communist party. He had just been discharged after two years in combat during the shooting war in Korea. He had entered as a buck private and left at that rank. When commanding officers read the fellow's file and saw that word *Communist*, they didn't dare to promote him, even as far as private first class.

I stopped by the office of a civilian physicist friend back at Fort Monmouth and told him about my encounter in Seattle. The man panicked. He closed his door, sat me down, and furiously told me never to mention that incident again. I would get him fired, and I shouldn't think that I was invulnerable, either.

He was right, as it turned out, although the army was not generating the fear. Whom I had once known was of little concern to them, as long as the word *Communist* did not turn up on my own record. My security clearances and my loyalty were never questioned by the military or by any subsequent government agency. The Red Scare was an expression of civilian—of public—fear. My past did not catch up with me until the mid-1960s, when I was teaching at Washington State University in Pullman, Washington, across the state from Seattle.

One day, the new chairman at the University of Washington drove over from Seattle to tell me that my friend Charlie Costello had died at the age of only 30 from infantile diabetes. Charlie and I had been major investigators of phase-change heat transfer in the Northwest.

The chairman asked me to come back to Seattle and take over the phase-change laboratory that Charlie had built. I can still feel the clawing discomfort of trying to digest Charlie's death and this fine opportunity— both at the same time. I said I would come, and the chairman went back to formalize an offer. A week later, he called, horribly embarrassed, to retract the offer. The All People's Student Center, and my former life, had resurfaced in the department.

However, this was a late, and pretty minor, aftershock of the Red Scare. By the spring of 1954, McCarthy had been humiliated during his fact-scrambled attack on the army, and he had finally turned upon President Eisenhower himself. That fall, the Senate finally censured him, and his purges began losing most of their bite.

McCarthy lived another three years while his alcoholism became increasingly severe. He made repeated trips to the Bethesda Naval Hospital to be dried out. His liver finally failed, and he died on May 2, 1957, at the age of only 48.

The same year McCarthy died, two manifestations of radical upheaval came alive—two manifestations that lay on opposite ends of any functional spectrum. One was the Beatnik revolution—a movement that deserves far more attention than it has received. The other was Sputnik.

First, the Beatnik movement. It was a literary movement, not yet overtly political. The hippies, the free-speech movement, and Vietnam protests all swallowed it up seven years later. They were radical revolution full-blown. But the beatniks had begun the upheaval more subtly, with a trick of etymology that is hard to explain.

Millions of GIs had come home in 1946, filled with a craving to reestablish domestic normalcy. They wanted a house with a white picket fence and three children. They wanted an eight-to-five job. They wanted normalcy, normalcy, normalcy. What they got soon cloyed into stultifying orthodoxy.

By the mid-fifties a slang term had found its way into a subculture of young people, increasingly jittery about the smallness of the dreams that surrounded them. A typical piece of dialog went like this: "Hey, man, how ya doin'?" "Dunno, man. I'm beat." What did *beat* mean? Tired? Beaten down? Ground up by the system? All of the above.

Soon people were speaking of the *Beat Generation.* Hemingway, that creature of *Modern,* had been called a member of the *Lost Generation.* The young of the 1920s had been lost in the seas of modern excess. The young of the fifties were now lost in a sea of moderation.

Writer Jack Kerouac came out of the mill town of Lowell, Massachusetts, and reacted. He was an athlete—tall, handsome, all-American, fundamentally shy, and clear in his mind that his generation would have to transcend the ennui. Kerouac created a quasi-autobiographical literary counterculture, a world in which the disenchanted regained their enchantment. He puts these words in the mouth of his fictional friend Dean in *On the Road*: "Man, wow, there's so many things to do, so many things to write! How to even begin to get it all down and without modified restraints and all hung-up on like literary inhibitions and grammatical fears . . ."[6]

That makes a very accurate account of how Kerouac created the book in 1951. He typed a hundred words per minute, and to avoid being slowed down he taped paper together and did the first draft on a single scroll of paper without stopping to change the pages in his typewriter.[7]

On the Road tells of his great looping journey across America, a journey in which he opens his senses to the land and to the people. Kerouac's generation would be the blessed ones, the ones who would not play the game, the ones to check out of the gray-flannel world around them and, instead, seek beauty indiscriminately—in cracked plaster, in the grunge of the city streets, in hitchhiking across Nebraska. In Kerouac's world, *beat* was transmuted into shorthand for *the beatified*. His was the generation of the blessed ones, the ones who would see God in all things. Kerouac became a Buddhist.

Kerouac's friend, poet Allen Ginsberg, was four years younger than he, and Ginsberg radicalized these themes. His poem *Howl*, published in 1955, describes Kerouac's beaten world in far more dire terms. He says of the best minds of his generation, "What sphinx of cement and aluminum bashed open their skulls and ate up their brains and imagination?"[8]

I fell into this world in the spring of 1956, soon after the army had finished with me. I was an untenured instructor, on a one-year appointment at the University of Washington, teaching machine design and hanging out with the poetry students of Theodore Roethke and Stanley Kunitz. One night a young, clean-shaven Ginsberg showed up at a house party. He had just ridden the bus up from Berkeley. I fell into a long conversation with Ginsberg and a smart physics student. In the course of it, those two virtually defined the terms of our fragmenting world.

The student tried to catch Ginsberg in a straightjacket of mathematical logic, while Ginsberg dodged and weaved. Logic was one more marker of the orthodoxy that Ginsberg meant to slip clear of. On the way up to Seattle, he had penned his own version of modern physics while he contemplated the luggage racks in the bus:

> The racks were created to hang our possessions, to keep us together,
>   a temporary shift in space,
> God's only way of building the rickety structure of Time.[9]

Yet, much of Ginsberg's rejection of logic was a kind of role-playing. A University of Houston colleague tells about a meeting she attended, years later. A young man was lecturing on the free form of Ginsberg's *Howl*, while a scruffy older man in the seat next to her grew increasingly agitated. Finally the old man leapt to his feet, interrupted the speaker, and said, "Listen here. I sat down and studied Walt Whitman's metrics for six months before I wrote *Howl*."[10] *Howl* was so powerful just *because* Ginsberg had shaped its seemingly wild words with a very disciplined hand.

Nine months after that party, I had moved down to Berkeley, and there I found the whole world seeming to converge upon the Bay area. Ginsberg was now there, forming the epicenter of the new literary movement. My wife-to-be was living in San Francisco, and, on our third date we went to a poetry reading at the same Six Gallery where Ginsberg had first read *Howl*. Photographer Jerry Stoll caught us there, and later included our photo in his fine pictorial account of the Beat revolution, *I Am a Lover*.[11] The caption for the photo, a quotation from Franklin D. Roosevelt, is a sly reference to the rumble of impending change going on around us all. It says:

> This generation of Americans has a rendezvous with destiny.

Perhaps we did. But revolution would not really reach the Berkeley side of the bay until long after we had left. For the moment, Berkeley was still, in the eyes of San Francisco, the land of the gray-flannel suit—conservative and orthodox. One of the last things I remember from Berkeley in 1961 was a chance meeting with a young musician friend, and another girl who was introduced to us as the daughter of civil engineering professor Hans Albert Einstein (Albert Einstein's granddaughter.) We made chitchat and then asked what they were up to.

"Oh, we're going over to San Francisco to join a protest against the House Un-American Activities Committee. Would you like to come along?" We didn't go. The revolution still seemed to belong at the printing press, not in the Place de la Bastille. Besides, the revolution was still rising silently, under our feet, like a great tsunami that had yet to reach the shore.

Not until 1964 did the Beat movement reach the other side of the bay, fully mutated into overt political action with sit-ins, protest marches, and many other forms of communal rebellion. For the moment, it was simply called "The San Francisco Renaissance"; and it was still being shaped by individuals, not by mobs.

Yet it was all around us, and it was being shaped in many ways. The Beats formed an important piece of it, but far from the only one. At the same time that the San Francisco Renaissance was emerging in spring of 1957, I went to what was, for me, a very important lecture on the Berkeley campus.

A young black minister, only one year older than I, had just organized a protest against segregation on the buses of Montgomery, Alabama. He was, of course, Martin Luther King, Jr., and his talk was riveting. In the end he said an odd thing, something that has often been lost in King's legacy: "Our aim is as much to deliver white people from the wrongs of segregation, as it is to deliver black people. If we forget that, then we've failed." That remark places King in the civil rights revolution much in the way Kerouac's concept of beatitude places him in the Beat revolution. For revolution always occurs in two stages: the moderates begin it; the radicals finish it.

King was followed by James Baldwin, and James Baldwin by Stokely Carmichael—just as surely as Kerouac was followed by Ginsberg, and Ginsberg was followed by Mario Savio. The 1950s revolution was, in fact, a complex of revolutions, and all of them evolved in this way.

Take, for example, music: If *Modern* began with people like Schönberg gutting the hierarchy of tonality, by the 1950s, composers like John Cage were now out to remove every remaining constraint. The Berkeley music department did a major performance of Cage works in 1959. Two pieces on that program give a sense of what he was doing.

In Cage's work for 12 portable radios and 24 operators, *Imaginary Landscape no.4*, 12 people reset the knobs to different stations according to his score. The other 12 varied the volume. (My wife played one of the volume controls.) The performance took place in a new, steel-framed performance hall, and what little sound emerged was only static. Far from being bothered by this twist, Cage was delighted.

The other piece was for piano and titled 4'33". A pianist sat down, opened the piano lid, cracked his knuckles, punched a stopwatch, and then observed four minutes and 33 seconds of silence, as we listened intently. The moment was broken only by one outraged attendee who got up, muttered loudly, "I've had enough of this shit," and walked out. (In 2002, Cage's heirs sued composer Mike Batt for including a piece titled *One Minute of Silence* on a new album. It was widely reported that they won a six-figure settlement, and Batt was widely quoted as saying, "Mine was a much better silent piece. I have been able to say in one minute what Cage could only say in four minutes and 33 seconds.)"

Modern thus reached its reductio ad absurdum in many ways. A continuous epoch of invention, and improvement of our standard of living, had been running for two generations by the late fifties. Now it was collapsing. Invention would continue, as would the improvement of our standard of living. But we would now have to develop a wholly new metaphorical language in which to think about change. We needed a sign from heaven, and we received it on October 4, 1957.

The Russians and Americans had both been racing to put a satellite into orbit. The Russians got there first with Sputnik I, a 183-pound spherical satellite about the size of a basketball. Sputnik's A-flat, beep-beep-beep radio signal sent shockwaves through a society that knew it was floundering.

Our reaction was odd. We collectively drew ourselves up and said: "Those Russians! They know how to build a big beefy booster rocket, but we Americans are far more sophisticated. We'll use better science to make smaller and more versatile satellites." That may not have been entirely hubris, but a number of serious failures still lay ahead in our own space program.

The immediate result of Sputnik was in direct keeping with that response, however. If you were an engineer, you now wanted to become an engineering scientist. If you were in engineering science, you would become a physicist. Physicists wanted to become mathematicians, and many mathematicians began to look like philosophers. Where the philosophers went, I am not certain.

However, we all tried to be more sophisticated than we had been. Grade schools began introducing children to set theory, graduate school enrollments went up, and every engineering faculty member now had to have federal research grants.

The odd thing is that the space program was a continuation of *Modern*. Indeed, one rocketeer from the early days of *Modern*, Werner von Braun, was now a primary player. The space program had been dancing across the pages of *Popular Mechanics* for decades. Rocketry would remain strong until about 1971, and would then be ratcheted back to secondary status for a long time. *Modern* had to give way to new directions entirely.

We had entered World War II using electromechanical analog computers. But in 1937 a young physics instructor at the University of Iowa, John Atanasoff, invented the digital computer. He realized that a machine could easily manipulate simple on-off electrical pulses. By doing computations in the "either-or" number base of two, instead of in base ten, a machine could execute calculations naturally and very rapidly.

During World War II the military developed the first working digital computers using fragile radio tubes in their logic circuits. Those computers were huge, isolated machines. In 1943, Thomas Watson, chairman of IBM, made his now-famous remark, "I think there's a world market for maybe five computers."

And, indeed, to own one of those machines was to own the fabled white elephant, because, with thousands of tubes, burning out with tedious regularity, a man in a white coat had to be constantly at work. But this was the way we saw the new computers, and no one paid much attention in 1952 when British scientist G.W.A. Dummar wrote: "It seems now possible to envisage electronic equipment in a solid block with no connecting wires. The block may consist of layers of insulating, conducting, rectifying and amplifying materials, [and] electrical junctions."[12]

Dummar's idea of casting a set of electronic functions into one monolithic electric element stood to vastly reduce the rate of failures. In July 1958, Jack Kilby of Texas Instruments, now able to replace the old tubes with transistors, finally created such an integrated circuit. A few months later, Robert Noyce, head of Fairchild Semiconductor Corporation, created a slightly better version, independently, and a patent war was underway.

After dumping money into the courts for years, Fairchild and TI saw how foolish their combat was. They agreed to forget the lawyers and share the idea. Kilby and Noyce acknowledged each other's contributions, and life went on. It was a very wise thing to do. By 1969 both Fairchild and TI had managed to put complete central processing units on single chips. Then Noyce formed a new company, INTEL, for INTegrated ELectronics, and he started producing whole computer motherboards. Computer costs plunged; but, even then, we didn't see where all this was going.

In 1977, the president of Digital Equipment Company could still say, "There's no reason people would want computers in their homes." And he might have been right, but technological futures are always unreadable. Who could have predicted that the early 1980s would bring shelf upon shelf of shrink-wrapped commercial packages of software into new computer stores? Once you and I were able to use our computers without writing the programs, computers *did* enter our homes. And they entered the closest quarters of our daily lives as well.

My work at Fort Monmouth had been designing mechanical equipment for both processing the new transistor materials and measuring their properties. We knew the work would be useful in many applications, but we had no idea that the transistor would revolutionize every aspect of the world we lived in. We were still focused upon the metaphors that made *Modern*. We were asking such questions as, "How will the transistor impact the airplane and the automobile?"

It took time for us to weave new metaphors and new icons around the fruit of that work. Kerouac had continued to use the *automobile* as a primary metaphor. Around it, he created a subterranean countercultural network, woven through America. Yet the automobile was still the same icon that had helped to define *Modern* in the 1920s and 1930s.

The ultimate icon of retrograde America as we entered the 1960s was, oddly enough, the staid, formal, IBM computer engineer visiting companies and universities in his blue suit and dark necktie. The computer was a presence, but it had yet to shape any metaphor for living in the twentieth century. It was not *Modern;* it was not a replacement for *Modern;* and it was certainly no visible part of any revolution. Nor would it be, until it finally fell into the hands of the public in the 1980s.

Modern was such a powerful epoch that we have not *yet* figured out what will replace it in our lives. What framework of metaphor and machine can energize us once more, as my parents were energized? We cannot go back, nor do we want to go back. But Kenneth Clark concluded his *Civilisation* series with this remark: "I said at the beginning that it is lack of confidence, more than anything else, that kills a civilization. We can destroy ourselves by cynicism and disillusion, just as effectively as by bombs."[13] Clark wrote that a decade after the dog days of *Modern*, and a general tone of pessimism attended his conclusion. He expressed serious concern about the sundering of confidence that he saw going on in the wake of *Modern*.

Thus our task in the next and last chapter will be to ask whether or not we postmodern people, struggling to regain our faltering traction, have yet done so and, if we have not, just how we eventually will.

16

After *Modern*

odern was not displaced by other ideas. Rather, it simply ran out of energy. Technological growth did not abate, nor did the seemingly new technologies that arose in the near wake of *Modern*—the space program and computers, for example—seem discontinuous with the past. Rocketry and computers could easily have lent themselves to the unique verve of *Modern*, and they had, indeed, been evolving for a long time as a part of *Modern*.

I might liken the end of *Modern* to a great caravan that reaches one more river, perhaps a little more formidable than most of the previous rivers it has crossed. There the travelers finally lose their collective confidence and stop. Others will pick up the journey on the other side, but they will do so without the old élan. Others will move forward, just as rapidly, but now questioning themselves, wondering what their journey is truly worth and no longer deriving the same primal joy from it.

We can see how this worked if we visit the last great exemplar of *Modern*. He was Vannevar Bush, America's leading government counselor on matters of technology and science from before World War II until 1959. Bush fit the profile of *Modern* perfectly. Born in 1890, he studied electrical engineering and mathematics at Tufts University and MIT. He was made the chairman of the NACA, the National Advisory Committee for Aeronautics, in 1938. (That was the organization that President Eisenhower recreated as NASA on July 29, 1958.) Bush continued in that post until 1941 but was meanwhile appointed the chairman of the president's National Defense Research Committee in 1940.

In 1941 Bush was made director of the Office of Scientific Research and Development—a post he held until 1947. For two decades after World War II he served in high administrative posts in both education and government. He served as president of the Carnegie Foundation and chairman of the MIT Corporation. In 1950, Bush founded the National Science Foundation.

Bush's biographer, G. Pascal Zachary, likens him to Benjamin Franklin in several respects.[1] Bush, like Franklin, was enormously creative. Like Franklin, he had an astonishing ability to pick up the right threads of technological change. And, like Franklin, he set out to use politics as the means for making technology serve us in the best possible way.

Bush certainly represented *Modern* in his belief that the purpose of technology was to reshape the texture of human life on Earth. Yet, by the mid- to late 1950s, he realized that he had reached the water's edge. He knew that, once he crossed over (*if* he crossed over) he would be embarked upon a different journey. He foresaw several areas in which we were on the verge of major changes in our course. But, in the end, he was unable to negotiate the changes himself. Let us look at three examples of the poignant way that dilemma played out in his life.

First, if Bush shared that prime tenet of *Modern*—that technology could take us to the moon—he did so only figuratively. He was, in fact, sharply critical of the speed with which the government was pursuing the space program. That may strike us as shortsighted today. But, as it turned out, the space program did indeed founder as soon as it had made its leap to the moon.

Bush's reason for opposing an accelerated space program was distinctly unmodern. He believed that it was too dangerous; he feared that astronauts would die. He was no longer thinking like the *savage boy inventor* of the early twentieth century. And here the plot thickens: The latest shuttle disaster makes it clear that space travel is still dangerous after fifty years. We could not dodge the danger in Vannevar Bush's time and we cannot dodge it now. Yet, *despite* a reawakened awareness of danger (or maybe *because of* it) I think the year 2003 brings us close to seeing the space program re-energized (more on that in a moment).

The second example is Bush's relationship to the red scares of the early fifties. Bush was pretty conservative when those assaults began, and, at first, he figured that America really did need to purge itself of communist sympathizers. Indeed, he had felt that Einstein was too left-leaning and had joined in blocking him from active involvement in developing the atom bomb during the war.

It was only in 1953, when Bush's highly regarded friend J. Robert Oppenheimer came under attack, that Bush was transformed. The main reason for the attack on Oppenheimer had been his opposition to developing a hydrogen bomb—an opposition that Bush had shared. Bush knew Oppenheimer very well, and, at that point, he finally understood that the world had gone crazy. After that, Bush slipped into despondency and his light faded. Zachary writes, "The Oppenheimer affair certainly high-lighted the gulf between him and the national-security elite. Having been 'present at the creation,' as Dean Acheson put it, Bush now watched a new American nation journey in what he saw as the wrong direction."[2] Zachary goes on to call him now a "hero without a cause," who in his late years "seemed to be against *everything*." When he was 65, Bush began a slow retreat from much of his public life.

The third example of Bush's realization that a dramatic shift of em-phasis was in the wind came earlier. And, despite one gaping limitation, it was a remarkable perception. During the last days of World War II he published an *Atlantic Monthly* article that offered an unalloyed predic-tion of future technology.[3]

This time Bush was remarkably on target in the big picture, while his misunderstanding of one crucial detail reveals the way he failed to see across that river that divided *Modern* from post-*Modern*. During the 1930s, Bush had taken the analog computer about as far as anyone could. His Rockefeller Differential Analyzer was a huge machine that occupied several rooms at MIT and used electromechanical elements to approxi-mate calculations. It was the grandest computer ever built.[4] Yet, by 1945, the new digital computers had made it obsolete.

The analog/digital issue is the only thing that mars Bush's otherwise stunning prediction, and it may seem that I am picking a nit even to mention it. This, however, was the fork in the road—the point at which the world shifted upon its axis.

We use the word *analog* for calculations that are analogous to the continuous processes we experience in nature. Speaking technically, we might, for example, create an electrical circuit that yields a cumulative electric charge. We then arrange that circuit to make the electric charge accumulate in the same way that a particular mathematical integration accumulates a sum. A digital computer is more abstract. It breaks text, or any other kind of information, into sequences of on/off signals.

The central issue that drove Bush's vision was his search for ways to cope with the mounting flood of information. The flood was only going to get worse, and he knew that effective handling of information would have a huge impact on both future technology and the very texture of human life. He therefore imagined a scheme that he called *memex*. He described it thus: "Consider a future device for individual use, which is a sort of mechanized private file and library. It needs a name, and, to coin one at random, 'memex' will do. A memex is a device in which an indi-vidual stores all his books, records, and communications, and which is mechanized so that it may be consulted with exceeding speed and flex-ibility. It is an enlarged intimate supplement to his memory."[5]

Bush's memex system was to have used microphotography and optical scanners, aided by the analog computer. He never whispers the words "digital storage." The extent of storage capacity that he conceives has, consequently, been surpassed long since. It has expanded to contain not just our private knowledge, but that of the whole planet as well. The Internet has vastly outstripped memex.

The computer was the lynchpin of his prediction. He saw with perfect clarity that it was going to permeate information handling in everyday life. For example, his article takes the reader to a new department store. A customer hands the clerk his private punched card. The clerk adds a price card from each item. He lines them up in a slot. The machine automatically corrects the inventory, charges the customer, and prints a receipt.

These were the clear antecedents of the credit card and the barcode, but Bush saw them both in analog form. He also wanted to encode speech, and here he grew vague on how we might do so. That is hardly surprising, since practical sound storage and retrieval systems had to wait until on/off digital sound storage was available to do it.

While the other futurists of 1945 turned their crystal balls on travel and power production, Vannevar Bush clearly saw that the next great intellectual frontier would lie in managing information. He finished this remarkable essay with a point that has yet to be tested, despite all our stunning information storage and retrieval capacities. He said:

> Presumably man's spirit should be elevated if he can better review his shady past and analyze more completely and objectively his present problems. He has built a civilization so complex that he needs to mechanize his records more fully if he is to push his experiment to its logical conclusion and not merely become bogged down part way there by overtaxing his limited memory. His excursions may be more enjoyable if he can reacquire the privilege of forgetting the manifold things he does not need to have immediately at hand, with some assurance that he can find them again if they prove important.
>
> The applications of science have built man a well-supplied house, and are teaching him to live healthily therein. They have enabled him to throw masses of people against one another with cruel weapons. They may yet allow him truly to encompass the great record and to grow in the wisdom of race experience.[6]

Will we be an altered and wiser people, now that we have both achieved and exceeded Bush's goal? I would be happier if I saw better evidence of it. The weak link in the information chain is our inability to hold more than a tiny whisper of that cornucopia of knowledge in our own frontal lobes at any moment. We need more than just access to *stored* knowledge.

In any case, Bush saw what other prophets did not. He recognized that the twin issues of information retention and access would dominate the end of the twentieth century. Yet, he remained modern. For the very tone of his words—the expressions of hope—were remarkably similar to those we had been reading for two generations in *Scientific American* and *Popular Mechanics*.

One new science/technology in particular underlay the movement into the information management domain; and, more than any other, it determined the shape and form of the era that followed *Modern*. By making digital signal management natural and simple, the *transistor* was about to redirect human energy. Its first fruit was the truly ultra-light battery-operated portable radio, first popularized by Sony. Suddenly we carried in our pockets what had, only a decade before, been a primary piece of living room furniture.

By the 1960s, transistors were also serving the new medium of television. TVs had started becoming commonplace in American homes during the 1950s. In 1960, television first determined the outcome of a presidential campaign. Kennedy's people had figured out how to use it, while Nixon's people had not. In 1963, Kennedy was shot, and Americans spent a week doing nothing but watching their still-new television sets. By the mid-sixties we all knew that TV would henceforth be, not an amusement, but a major force in our lives.

The television set, however, retained as its primary element the cathode ray tube, one of the long-surviving electronic tubes—still kin to radio tubes. Television sets also used myriad radio-type tubes in their supporting circuits until the 1960s. After Sony had rid its small radios of tubes, the tubes in radios and television sets everywhere were fairly quickly replaced with more durable, reliable, and energy-efficient transistors. But only as we enter the twenty-first century do we see the old CRT tubes in our TV sets and computer monitors giving way, at last, to purely digital displays.

In the new digital computers, transistors did more than just replace their diode tubes. They also made possible the creation of the integrated circuit, which became commonplace in the 1970s. Once that happened, the transistor could be credited with bringing about information handling on a scale far greater than Bush had ever dreamt. But it also brought on consequences no one could possibly have predicted. Our early reaction to the vastly enhanced information access has been unintended and unexpected. It has been *disillusionment*.

So much of what we learn is negative and has made us cynical. People have always been manipulated to some extent by politicians and by corporate interests, but now we are flooded with examples. More often than it surprises and delights us, new technology now makes us ask, "What will go wrong this time?" We so drown in information that we cannot see the good news.

And good news is definitely there. Human life expectancies have improved remarkably since 1950. Our medical capabilities have become downright stunning. Cancer was a death sentence in 1950. Today it is a threat and a damn nuisance. Tomorrow we can well hope that it will be gone. Genetic engineering promises enormous hope for the world's still-all-too-numerous hungry people, and it has already fulfilled much of that promise. But (unless we ourselves are hungry) we see its hypothetical dangers far more clearly than its palpable benefits.

Only one other time has people's access to knowledge taken the kind of leap forward that it did in the latter twentieth century. That was in the wake of Gutenberg's creation of a workable system for printing books with moveable metal type in 1455. By 1501, an astonishing flood of new printed books had suddenly been created. Something between eight and 24 million volumes—some 30,000 titles—had suddenly arrived in a world where a well-stocked library might previously have held only 20 or 30 books.[7]

The result was extremely complex. By the early sixteenth century, people reeled under the assault of all that information. Any in-depth look at the Renaissance would be out of place here, but the early results were really quite muddled. There followed a great leap forward in art, science, literature, and architecture—all accompanied by widespread social upheaval.

The Protestant reformation became violent. Witch burnings took a dramatic upturn, especially in the Protestant countries in which printing had begun. Women lost much of the fairly strong position they had enjoyed in medieval society. The countries of Western Europe began practicing slavery. Gunpowder came of age and war became more horrible than ever.

One image from the early Renaissance and the near wake of Gutenberg is very telling. It was provided by Albrecht Dürer in 1514. Dürer was more than just a fine artist, he was also a shrewd observer of his times and a very effective user of the new technology of print. Among the huge array of images that he left us, this one was particularly disturbing, and it has drawn vast attention from art historians.

The image is one of the new copperplate engravings—a technology that Dürer helped to perfect. Its title is *Melencolia I,* and, in it, he shows us an angel empowered with intelligence, ability, and even wings with which to fly. This glorious, better-than-human creature sits, despondent, among all the medieval scholar's instruments of natural philosophy, mathematics, and technology—all the symbols of the old alchemy and of the neoplatonist world that was now crumbling in the face of new ways of thinking.

Dürer shows what has become of the very world that his science and art had so helped on its way to devastating change. He engraves it upon

Melencolia I, by Albrecht Dürer, 1514.[8]

the face of this beautiful being. And he was not alone in recognizing that things would never be the same. His contemporary and friend, the scholar Erasmus, uttered a somewhat more optimistic reading of ultimate change only three years after Dürer drew *Melencolia*. Yet, in Erasmus's famous cry, "Immortal God, what a world I see dawning. Why can I not be young again?" we hear the same anguished recognition of irreversible upheaval.

Though separated by over four centuries, Dürer and Bush were clear in their minds that they stood on thresholds of new eras in communications, and that massive changes would attend those eras. Defeat began enshrouding Bush after the late fifties. And, although Dürer remained highly productive right up to his death in 1528, he sounds a chilling note of warning in this remarkable and disturbing engraving.

We have been in our own state of melancholia for over four decades now. Surrounded by plenty, we have watched technology taking us from one height to the next. Yet we (in a most unmodern way) have lived in constant fear of our own creative cornucopia.

We, like Dürer's melancholic angel, are standing in need of what I can only identify as spiritual regeneration. And I think that it awaits us. We have spent almost a half century regrouping, and now many straws are in the wind. Let us look at just two ways in which that regeneration of human energy might be brought about.

The first is in space exploration: On Christmas Eve, 1968, astronaut Frank Borman rode his orbiter around the far side of the moon. He was the first living creature to gaze on that surface, and, as he came around the moon, he read aloud from Genesis: "In the beginning God created the heaven and the earth." The vast expanse of space seemed to open up in that moment. Suddenly the firmament was no longer an abstraction. Hope was immense.

A few months later, we all saw that small footprint upon the dust of the moon. It seemed to tell us that we had completed what we had set out to do, and the culmination did not match our dreams. Something Thoreau said rather mirrors the way we reacted to the space program at that point: "The youth gets together his materials to build a bridge to the moon or perchance a palace or temple on the earth—& at length, the middle-aged man concludes to build a woodshed with them."[9]

Now, however, we might actually build that figurative bridge. A number of developments point to a revitalization of our move into space. The great impediment to space travel has been the enormous cost of getting free of Earth's gravity. The three-stage rocket offers very inefficient means for doing that. However, as engineers have developed remarkably light and strong new materials, far less costly single-stage recoverable rockets are practically within reach.

An even more remarkable technology for leaving the clutches of Earth's gravity stands waiting in the wings. In 1979, Arthur C. Clarke wrote a science-fiction book, *The Fountains of Paradise.*[10] A few months later, he wrote a two-page letter which you can read at the end of Charles Sheffield's science-fiction novel, *The Web Between the Worlds.*[11] The letter absolves Sheffield of plagiarism. You see, Sheffield had cooked up the same mad premise that Clarke had.

Both authors proposed to replace rocket launches with an elevator, anchored at the equator and reaching up beyond the level of geosynchronous orbit, where an orbiting vehicle rides directly over one point on Earth. In his letter, Clarke congratulates Sheffield for arriving at the idea independently.

And if their idea sounds crazy, bear in mind that both Clarke and Sheffield were distinguished scientists who later turned to science fiction. Clarke was a noted astronomer. Sheffield was chief scientist of the Earth Satellite Corporation. Both used science fiction to teach us about space travel.

The main element in their elevator was a cable with an extremely high tensile strength—strong enough to support its 22,248-mile length. The cable was to be a single perfect crystalline carbon or silicon fiber. (Today engineers are closing in on carbon nanotube materials, which should provide even greater strength.) And we have routinely been putting geosynchronous satellites into orbit, ever since 1963.

Above the geosynchronous level, a satellite will fly away from earth; below it will fall back into Earth. The cable above would pull on the cable below and hold it in place. The cable would be tapered in thickness—thickest at the point of geosynchronous orbit, where the stresses would also be greatest.

I won't recount all the technology that Clarke and Sheffield describe to make such a thing work or how they plan to put such a thing in place, to anchor it, and to counterweight it above the geosynchronous level, or how they propose to pass the energy of rising and falling cars back and forth. But the two proposals are so similar because the laws of physics bind both authors. Others had actually invented Sheffield and Clarke's wild idea independently before them. A Russian engineer suggested a version, which he called a "Cosmic Funicular," as early as 1960.

One might incline to laugh this behemoth off as too grand to take seriously. However, at this writing, NASA has just sponsored a conference for the purpose of exploring design issues involved in actually building such a thing. The echoes of the Thoreau quotation about building a bridge to the moon are suddenly too strong to ignore.

Many such ideas have lain dormant for years and are only now being resurrected—photon propulsion, extended human dormancy during long space journeys, and so forth. However, what has really been dormant is not scattered *elements* of a space program, but a serious desire to get ourselves out there. I fully expect us to regain that desire, and to set off on a far grander scale than anything we have yet seen or even seriously contemplated. When we do, perhaps we can quit miring ourselves into the morass of self-consuming troubles that have created such melencolia here on Earth.

The other emergence that promises, once again, to kick-start human energy, is the fairly recent unfolding of radical new physics. Henry Adams heralded *Modern* when he told us we were facing the "irruption of forces totally new," and that we would have to rediscover "mystery." We did rediscover mystery in the reason-defying physics of quantum mechanics and relativity theory. And we have put all that to use.

Now we are just seeing a new yawning pit of implausibility opening up before us. Every day we read some new article about super-string theory, loop theory, dark matter, alternate universes, quantum computation, the cosmic anthropic principle, the implications of a highly articulated chaos theory upon our perception of reality, and simultaneity of communication between particles, all being backed up with new evidence.

For me to undertake a disquisition on any of these topics would be foolhardy beyond belief. For we stand on the threshold of *mystery* in the same way we did in 1900. Once more, our world seems about to become unrecognizably *larger* than it has been. Once more, hope wells up from that great ocean of untapped but inchoate *possibility*. Something is about to happen.

Modern is gone. It was such a wonderful epoch—joyous in so many ways, childish in others—but it is gone. Now we are about to become a new people, no longer *Modern* or postmodern, but something else entirely. We need a word for it.

We are about to become *Expanded*.

I like that. My parents were modern. My great-grandchildren will live in the *Expanded* era, that ancient time, back in the twenty-first century, when life once more became larger than life. As I think about it, I know precisely how Erasmus felt when he shouted at the confused world all around him, "Why can I not be young again?" Yet, though I know how he felt, I would not hunger to do it all again. For my time was *Modern,* and *Modern* was more than enough for any one person.

Notes

Chapter 1

1. M. Berman, *All That Is Solid Melts Into Air: The Experience of Modernity* (New York: Penguin Books, 1982, 1988).
2. L. White, Jr., *Medieval Religion and Technology* (Berkeley: University of California Press, 1978): "The Legacy of the Middle Ages in the American Wild West," pp. 105–120.
3. For the Shelley quote, see H. Jennings, *Pandaemonium: 1660–1886: The Coming of the Machine as Seen by Contemporary Observers* (New York: The Free Press, 1985): "Frankenstein," pp. 141–142.
4. M. Christman, *1846: Portrait of a Nation* (Washington, D.C.: Smithsonian Institution, 1996).
5. Johann Heinrich Lienhard's Odyssey is told in a lengthy journal, which has been translated into English and published in three books, each contiguous with the one before it. The first, dealing with the years up to 1846, is H. Lienhard, *New Worlds to Seek*, tr. Raymond J. Spahn, ed. John C. Abbot, foreword by John H. Lienhard IV (Carbondale: Southern Illinois University Press, 2000). The second deals with his overland crossing to Sutter's Fort: H. Lienhard, *From St. Louis to Sutter's Fort, 1846*, tr. and ed. E. G. and E. K. Gudde (Norman: University of Oklahoma Press, 1959). The third tells of his three-year stay with Sutter before and after the California gold rush: H. Lienhard, *A Pioneer at Sutter's Fort, 1846–1850: The Adventures of Heinrich Lienhard*, tr., ed., and ann. M. E. Wilbur (Los Angeles: The Calafia Society, 1941).
6. G. Grant, *The Story of the Ship* (New York: Mclaughlin Brothers, 1919).
7. The story of Shreve, Fulton, and the western riverboat is told in both L. C. Hunter, "The Heroic Theory of Invention," *Technology and Social Change in America*, ed. Edwin T. Layton, Jr. (New York: Harper & Row, Publishers, 1973), pp. 25–46; and J. T. Flexner, *Steamboats Come True: American Inventors in Action* (Boston: Little, Brown and Company, 1944/1978), chapter XXV.
8. *America Illustrated*, ed. J. David Williams (Boston: deWolfe, Fiske & Company, Pubs., 1883).
9. C. Dickens, *American Notes* (with three illustrations by Maurice Greiffenhagen) (London: Chapman & Hall; New York: Oxford University Press, [186–?])

10. Heber C. Kimball later became first counsel to Brigham Young. His (and Heinrich Lienhard's) house is now a showpiece in the Mormon reconstruction of Nauvoo.

11. R. C. Hayden, *Black American Inventors* (Reading, Mass.: Addison Wesley Publishing Company, 1972). For additional material see this excellent web site: http://www.princeton.edu/~mcbrown/display/rillieux_biography.html.

12. See note 6.

13. J. H. Kemble, "The Genesis of the Pacific Mail Steamship Company," *California Historical Society* XIII, 4 (1934): pp. 386–406. J. H. Kemble, "The First Steam Vessel to Navigate San Francisco Bay," *California Historical Society* XIV, 2 (1935): pp. 143–146.

14. See note 4.

15. E. Lurie, *Louis Agassiz: A life in Science* (Baltimore: The Johns Hopkins University Press, 1988).

16. S. J. Gould, *The Mismeasure of Man* (New York: W.W. Norton & Co., 1981).

17. B. Wallis, "Black Bodies, White Science: Louis Agassiz's Slave Daguerreotypes," *American Art* (summer 1995): pp. 38–61.

18. S. J. Gould, "Agassiz in the Galapagos," *Hen's Teeth and Horse's Toes* (New York: W.W. Norton & Company, 1980): chapter 8.

19. D. Lavender, *A Rocky Mountain Fantasy: Telluride, Colorado* (Telluride: San Miguel County Historical Society, undated).

20. For information on Fleischmann, see P. E. Palmquist, *Elizabeth Fleischmann: Pioneer X-Ray Photographer* (Berkeley, Calif.: Judah L. Magnes Museum, 1990). (In this exhibition catalog there is a discrepancy as to E. F.'s age. According to the text she was reported as being 15 years old in the 1880 census, but the article also says she was 38 years old when she died in 1905. I am guessing that the secondary report of the census is somehow incorrect, and I accept the age of 38 at death.) Fleischmann married Israel Julius Aschheim, grand secretary of the District No. 4 B'nai B'rith, in 1900. At that time she did something quite unheard of in 1900. She adopted the hyphenated last name Fleischmann-Aschheim.

21. N. Bolotin and C. Laing, *The Chicago World's Fair of 1893: The World's Columbian Exposition* (Washington, D.C: The Preservation Press, 1992).

22. *The Columbian Exposition Album* (Chicago: Rand, McNally & Company, 1803).

23. F. Howard, *Wilbur and Orville: A Biography of the Wright Brothers* (New York: Ballantine Books, 1987), chapter 1.

Chapter 2

1. J. Gleick, *CHAOS: Making a New Science* (New York: Penguin Books, 1987).

2. G. Grant, *The Story of the Ship* (New York: McLaughlin Brothers, 1919).

3. N. Dean, "The Brief, Swift Reign of the Clippers," *American Heritage of Invention & Technology* (fall 1989): pp. 48–55.

4. For more on Alfred Ely Beach and his subway, see: O. E. Allen, "New York's Secret Subway," *American Heritage of Invention & Technology* (winter 1997): pp. 44–48; B. Bobrick, *Labyrinths of Iron: Subways in History, Myth, Art, Technology, and War* (New York: Henry Holt and Company, 1981, 1986): chapter 6, "The Lamp and the Ring."

5. Much has been written about Nast. See, e.g., T. N. St. Hill, *Thomas Nast: Cartoons & Illustrations. 117 Works* (New York: Dover Publications, Inc., 1974); M. Keller, *The Art and Politics of Thomas Nast* (New York: Oxford University Press, 1968); or

A. B. Paine, *Th. Nast: His Period and His Pictures* (New York: The Macmillan Company, 1904).

6. For the Swinburne quote (and for its use in context) see: J. Wosk, *Breaking Frame: Technology and the Visual Arts in the Nineteenth Century* (New Brunswick, N.J: Rutgers University Press, 1992), chapter 5, "The Struggle for Legitimacy: Cast Iron."

7. *Great Industries of Great Britain*. Vol. I (London: Cassell Peter & Galpin, 1877–1880), frontispiece.

8. The vast literature on cast-iron architecture includes, for example: M.D.J. Coulter, *Texas: Cast-Iron Architecture, A Photographic Survey* (The Texas Historic Resources Fellowship, 1975–1976); W. K. Sturges, *The Origins of Cast Iron Architecture in America* (New York: Da Capo Press, 1970); *Baltimore's Cast-Iron Buildings and Architectural Ironwork*, J. D. Dilts and C. F. Black, eds. (Centreville, Md.: Tidewater Pubs, 1991); J. Gay, G. Stamp, *Cast Iron* (London: John Murray, 1985); M. Gayle, D. W. Look, and J. G. Waite, *Metals in America's Historic Buildings: Uses and Preservation Treatments* (Washington, D.C.: U.S. Govt. Printing Office, 1980); and R. Lister, *Decorative Cast Ironwork in Great Britain* (London: G. Bell and Sons, LTD, 1960).

9. For more on cast-iron façades in Soho, see: M. Gayle and R. Lynn, *A Walking Tour of Cast-Iron Architecture in SoHo* (New York: Pub. Center for Cultural Resources, 1983); and M. Gayle and E. V. Gillon, Jr., *Cast-Iron Architecture in New York* (New York: Dover Pubs., Inc. 1974).

10. See, e.g., *The Oxford Book of Children's Verse in America*, Donald Hall, ed. (New York: Oxford University Press, 1990).

11. R. Garcia y Robertson, "Failure of the Heavy Gun at Sea, 1898–1922," *Technology and Culture* (July 1987): pp. 539–557.

12. J. A. Tarr and S. J. Fenves, "The Greatest Bridge Never Built?" *American Heritage of Invention & Technology* (fall 1989): pp. 24–29.

13. *National Historic Mechanical Engineering Landmarks*, R. S. Hartenberg, ed. (New York: American Society of Mechanical Engineers, 1979).

14. *Modern Mechanism*, ed. Park Benjamin, supplementary volume to *Appleton's Cyclopaedia of Applied Mechanics* (New York: D. Appleton and Company, 1892), pp. 708–714.

15. R. Buckler, "A Hidden Wonder of the World," *Invention & Technology* (spring 1997): pp. 40–45.

16. L. F. Baum, *The Master Key, an electrical fairy tale founded upon the mysteries of electricity and the optimism of its devotees. It was written for boys, but others may read it*, ill. Fanny Y. Cory (Indianapolis: Bowen-Merrill Co., 1901). *The Master Key* may be read online at: http://www.selfknowledge.net/b/mstky10.htm .

17. "Shawnee oppose Oz theme park," National briefs, *Houston Chronicle* (Oct. 10, 2000).

18. H. Adams, *The Education of Henry Adams* (New York: The Heritage Press, 1918). See especially chapter 2.

Chapter 3

1. The paper in which Planck completed his quantum prediction of the spectral emissive power of a radiating blackbody was: M. Planck, "Entropie und Temperatur strahlender Wärme," *Annelen der Physik*, 4, No. 4 (1901): p. 553. (Actually, Planck wrote two important preliminary papers in 1900.) For a simple description of the prediction see: C. L. Tien and J. H. Lienhard, *Statistical Thermodynamics*, Rev. ed. (Washington, D.C.: Hemisphere Publishing Corporation, 1979).

2. Four good sources on Dalton, the atomic theory, and Dalton's own ideas about notation are: G. Greenway, *John Dalton and the Atom* (Ithaca, N.Y.: Cornell University Press, 1966); *John Dalton & the Progress of Science*, ed. D.S.L. Cardwell (New York: Manchester University Press, Barnes and Noble Inc., 1968); H. C. von Baeyer, "Nota Bene," *The Sciences* (January/February 1999): pp. 12–15; and J. Bronowski, *The Ascent of Man* (Boston: Little, Brown and Company, 1973), chapter 4.

3. For Boltzmann's story, see P. Coveney and R. Highfield, *The Arrow of Time: A Voyage Through Science to Solve Time's Greatest Mystery* (New York: Fawcett Columbine, 1990). Boltzmann's H-theorem (which deals with reversibility) is discussed by C. L. Tien and J. H. Lienhard, *op. cit.* (see especially section 12.3).

4. S. P. Thompson, *Light Visible and Invisible*, 2nd ed. (London: Macmillan and Co. Limited, 1928).

5. N. Knight, "'The New Light': X-rays and Medical Futurism," *Imagining Tomorrow: History, Technology, and the American Future*, ed. Joseph Corn (Cambridge, Mass.: MIT Press, 1986), chapter 1.

6. R. Herzig, "Removing Roots: 'North American Hiroshima Maidens' and the X Ray," *Technology and Culture* 40, No. 4 (October 1999): pp. 723–745.

7. A. F. Collins, *The Boy Scientist* (Boston: Lothrop, Lee & Shepard Co., 1925).

8. S. J. Reiser, *Medicine and the Reign of Technology* (New York: Cambridge University Press, 1978).

9. H. Adams, *The Education of Henry Adams* (New York: The Heritage Press, 1918). See especially chapter xxv.

10. J. McGovern, *The Fireside University of Modern Invention Discovery Industry and Art . . .*, rev. ed. (Chicago: Union Publishing House, 1902), p. 37.

11. For a fairly elementary discussion of Schrödinger's equation, see C. L. Tien and J. H. Lienhard, *op. cit*, chapter 4.

12. For a nice discussion of Schrödinger's cat, see J. D. Barrow and F. J. Tipler, "Quantum Mechanics and the Anthropic Principle," *The Anthropic Cosmological Principle* (New York: Oxford University Press, 1988), chapter 7.

13. P. Lansky and G. Perle, "Atonality," *The New Grove Dictionary of Music and Musicians*, Vol. 1, ed. Stanley Sadie (New York: Macmillan Publishers, Ltd., 1980), pp. 669–673.

Chapter 4

1. A. Wallace, *The Prodigy: A Biography of William James Sidis, America's Greatest Child Prodigy* (New York: E. P. Dutton, 1986).

2. W. J. Sidis, *The Animate and the Inanimate* (Boston: R. G. Badger, 1925).

3. D. H. Feldman with L. T. Goldsmith, *Nature's Gambit: Child Prodigies and the Development of Human Potential* (New York: Basic Books, Inc., Publications, 1986).

4. O. Mayr, *The Origins of Feedback Control* (Cambridge, Mass.: MIT Press, 1970).

5. N. Wiener, *The Human Use of Human Beings: Cybernetics and Society* (New York: Da Capo Press, 1988).

6. M. Berman, *All That Is Solid Melts Into Air: The Experience of Modernity* (New York: Penguin Books, 1982, 1988). See especially chapter 1.

7. For more on Goethe and Faust, see B. M. Katz, *Technology and Culture: A Historical Romance* (Stanford: The Portable Stanford Book Series, 1990). See especially pp. 100–101.

8. T. P. Hughes, "Thomas Alva Edison and the Rise of Electricity," *Technology in America: A History of Individuals and Ideas*, ed. C. W. Pursell (Cambridge, Mass.: MIT Press, 1986), chapter 11.

9. For more on Tesla, see M. Cheney, *Tesla: Man Out of Time* (New York: Bantam Doubleday Dell Publishing Group, Inc., 1981). See also J. H. Lienhard, *The Engines of Our Ingenuity: An Engineer Looks at Technology and Culture* (New York: Oxford University Press, 2000), chapters 3, 10.

10. T. C. Martin, "Tesla's Oscillator and Other Inventions," *The Century Magazine* XLIX (April 1895): pp. 916–933.

11. For an excellent popular account of Einstein's enduring interest in invention, see T. P. Hughes, "Einstein the Inventor," *American Heritage of Invention & Technology* 6, No. 3 (winter 1991): pp. 34–39.

12. R. Sorensen, "Thought Experiments," *American Scientist* (May/June 1991): pp. 250–263.

Chapter 5

1. See any of a number of good collegiate art history texts. See, for example, M. Stokstad, *Art History* (New York: Harry N. Abrams, Inc., 1995), chapters 27, 28; or *Gardner's Art Through the Ages*, ed. Horst de la Croix, Richard G. Tansey, and Diane Kirkpatrick (New York: Harcourt Brace Jovanovich, Publishers, 1991).

2. The three *Century Magazine* articles are: T. Roosevelt, "In the Cattle Country" (with 11 illustrations by F. Remington), *The Century Magazine* 35, No. 4 (February 1888): pp. 495–510; T. Roosevelt, "The Home Ranch" (with 12 illustrations by F. Remington), *The Century Magazine* 35, No. 5 (March 1888): pp. 655–669; and T. Roosevelt, "The Roundup" (with 10 illustrations by F. Remington), *The Century Magazine* 35, No. 6 (April 1888): pp. 849–867. All 33 of the Remington illustrations are posted at http://www.uh.edu/engines/rem.htm .

3. G. E. White, *The Eastern Establishment and the Western Experience: The West of Frederic Remington, Theodore Roosevelt, and Owen Wister* (New Haven: Yale University Press, 1968), chapter 5.

4. For the story of Teddy Roosevelt in Medora, see D. McCullough, *Brave Companions: Portraits in History* (New York: Simon & Schuster, 1992), chapter 4.

5. H. Adams, *The Education of Henry Adams: An Autobiography* (New York: The Heritage Press, 1918). See especially chapters XXV and XXVIII.

6. W. C. Howze, The Influence of Western Painting and Genre Painting on the Films of John Ford, Ph.D dissertation 8728674 (Austin: University of Texas, 1986).

7. B. M. Vorpahl, *Frederic Remington and the West: With the Eye of the Mind* (Austin: University of Texas Press, 1972), chapter 1.

8. Volumes have been written about Remington's art. I especially recommend: E. B. Neff (with Wynne H. Phelan), *Frederic Remington: The Hogg Brothers Collection of the Museum of Fine Arts, Houston* (Princeton, N.J.: Princeton University Press in association with the Museum of Fine Arts, Houston, Tex., 2000). For more on both his art and his biography, see E. Jussim, *Frederic Remington, the Camera & the Old West* (Fort Worth, Tex.: Amon Carter Museum, 1983); and A. P. Splete and M. D. Splete, *Frederic Remington—Selected Letters* (New York: Abbeville Press, 1988).

9. L. Adele, "Charles Dellschau, 1830–1923," *Spirited Journeys: Self-taught Artists of the Twentieth-Century* (Austin: University of Texas Press, 1997), pp. 40–47.

10. Much has been written about the history of photography. I would recommend B. Newhall, *The History of Photography* (New York: The Museum of Modern Art, 1964).

11. An excellent general source on early women photographers is N. Rosenblum, *A History of Women Photographers* (New York: Abbeville Press, 1994).

12. P. Daniel and R. Smock, *A Talent for Detail: The Photographs of Miss Frances Benjamin Johnston, 1889–1910* (New York: Harmony Books, 1974).
13. T. T. Heyman, *Anne Brigman: Pictorial Photographer/Pagan/Member of the Photo-Secession* (Exhibition. Oakland, Calif: The Oakland Museum Oakes Gallery, September 17–November 17, 1974).
14. *Camera Work*, ed. Alfred Stieglitz, June 1913 (New York: A. Stieglitz): Oct. 1911 and June 1913 (courtesy of Special Collections, University of Houston Art and Architecture Library).
15. For more on the Armory Art Show, see Newhall, chapter 8; M. W. Brown, *The Story of the Armory Show* (New York: The Hirshhorn Foundation, 1963); and M. Green, *New York 1913: The Armory Show and the Paterson Strike Pageant* (New York: Collier Books, 1988).
16. R. M. Fogelson, *America's Armories: Architecture, Society, and Public Order* (Cambridge, Mass.: Harvard University Press, 1989).
17. See, e.g., *Gardner's Art Through the Ages*, pp. 1026–1027.
18. For more on Charles and Ray Eames, see, e.g., J. Neuhart, M. Neuhart, and R. Eames, *Eames Design* (New York: Harry N. Abrams, Inc., 1989). Although Ray Eames was listed as a co-author, this was published after her death and Charles Eames's death as well.
19. J. M. Marter, "The Engineer Behind Calder's Art," *Mechanical Engineering* (December 1998): pp. 52–57.
20. T. S. Eliot, *Four Quartets* (New York: Harcourt, Brace and Company, 1943).
21. S. Kern, *The Culture of Time and Space: 1880–1918* (Cambridge, Mass.: Harvard University Press, 1983) Quartet No. 4, "Little Gidding."
22. *Remington: The Years of Critical Acclaim*, ed. Kellie Keto (Fort Worth, Tex.: Amon Carter Museum, 1998), p. 1.
23. R. Cortissoz, "Frederic Remington: A Painter of American Life," *Scribner's Magazine* (February 1910): pp. 181–195.

Chapter 6

1. P. Pernin, *The Great Peshtigo Fire: An Eyewitness Account*, 2nd ed. (Madison: The University of Wisconsin Press, 1999).
2. See *New York Times* articles for the two weeks following Oct. 8, 1971.
3. W. Cronon, *Nature's Metropolis: Chicago and the Great West* (New York: W. W. Norton & Company, 1991).
4. C. Smith, *Urban Disorder and the Shape of Belief* (Chicago: University of Chicago Press, 1995), part I, "Fire."
5. P. Belluck, "Barn Door Reopened on Fire After Legend Has Escaped." *New York Times National Report*, Aug. 17, 1997, p. 10.
6. See, for example, T. Benson, *The Timber-Frame Home: Design, Construction, Finishing* (Newtown, Conn.: The Taunton Press, 1988).
7. T. Cavenaugh, "Balloon Houses: The Original Aspects of Conventional Wood-Frame Construction Re-examined," *Journal of Architectural Education* (September 1997): pp. 5–15. For a contemporary look at the Chicago balloon frame, see G. E. Woodward, *Woodward's Country Homes* (New York: Geo. E. Woodward, 1865). For more on the balloon frame, see F. W. Peterson, *Homes in the Heartland: Balloon Frame Farmhouses of the Upper Midwest, 1850–1920* (Lawrence: University of Kansas Press, 1992).

8. M. Culbertson, *Texas Houses Built by the Book: The Use of Published Designs, 1850–1925* (College Station: Texas A&M University Press, 1999). See also Culbertson's extensive references on mail-order houses.
9. G. R. Strakosch, *Vertical Transportation: Elevators and Escalators* (New York: John Wiley & Sons Inc., 1967).
10. *The Wonder Book of Knowledge*, ed. Henry Chase Hill (Philadelphia: The John C. Winston Co., 1923), pp. 232–241.
11. *Modern Mechanism*, ed. Park Benjamin (New York: D. Appleton and Company, 1892). See "Elevators."
12. R. M. Vogel, *A Museum Case Study: The Acquisition of a Small Residential Hydraulic Elevator* (Washington, D.C.: Smithsonian Institution, 1988).
13. P. Goldberger, *The Skyscraper* (New York: Alfred A. Knopf, 1981). The quotation is to be found on page 18.
14. T. F. Peters, "The Rise of the Skyscraper from the Ashes of Chicago," *American Heritage of Invention & Technology* (Fall 1987): pp. 14–23.
15. Texans might wonder about the San Jacinto Monument, which stands 12 feet taller than the Washington Monument. Finished 54 years later, it is not masonry, but steel-reinforced concrete—a hybrid between masonry and steel-frame construction. Consequently, it has only about two-fifths the weight of a corresponding masonry structure. It does correctly boast being the tallest "monument column" in the world.
16. "Liberty Enlightening the World," *Scientific American* LII, no. 24 (June 13, 1885).
17. R. M. Vogel, "Elevator Systems of the Eiffel Tower, 1889," *United States National Museum Bulletin* 228 (Washington, D.C.: Smithsonian Institution, 1961).
18. W. Worthington, Jr., "Early Risers," *American Heritage of Invention & Technology* (winter 1989): pp. 40–44.

Chapter 7

1. C. Jencks, *Skyscrapers-Skyprickers-Skycities* (New York: Rizzoli International Publications, Inc., 1980).
2. K. C. Gillette, *The Human Drift* (Boston: New Era Publishing Co., 1894). This was the first of several books in which Gillette set down his ideas. (The dedication reads, "The thoughts herein contained are dedicated to all mankind; for to all the hope of escape from an environment of injustice, poverty, and crime, is equally desirable.")
3. J. Mansfield, "The Razor King," *American Heritage of Invention & Technology* (spring 1992): pp. 40–46.
4. K. C. Gillette, *The People's Corporation* (New York: Boni and Liveright Publishers, 1924). (This was Gillette's last book, 30 years after the first. Here, the dedication says simply: "To MANKIND.")
5. See note 3.
6. C. Mierop, *Skyscrapers: Higher and Higher* (Brussels: Norma Editions, 1995).
7. See, for example, N. Brosterman, *Out of Time: Designs for the Twentieth-Century Future* (New York: Harry N. Abrams, Inc., Pubs., 2000).
8. H. H. Suplee, "The Elevated Sidewalk," *Scientific American* (July 26, 1913): p. 67 and cover picture.
9. C. Willis, "The Titan City," *American Heritage of Invention & Technology* 2, No. 2 (fall 1986): pp. 44–49.
10. Arie van de Lemme, *A Guide to Art Deco Style* (Secaucus, N.J.: Chartwell Books, Inc., 1986).

11. N. Messler, *The Art Deco Skyscraper in New York* (New York: Peter Lang Publishing, Inc., 1986).
12. C. Roth Pierpont, "The Silver Spire," *The New Yorker* (November 19, 2002): pp. 74–91.
13. S. Fox, "I Like to Build Things," *American Heritage of Invention & Technology* 15, No. 4 (summer 1999): pp. 20–30.
14. H. G. Wells, *When the Sleeper Wakes* (London: Harper & Brothers Publishers, 1899).

Chapter 8

1. For a look at this collection, see: *The Carriage Collection* (Stony Brook, N.Y.: The Museums at Stony Brook, 1986) (no author given); *19th Century American Carriages: Their Manufacture, Decoration and Use* (Stony Brook, N.Y.: The Museums at Stony Brook, 1987) (no author given); and *The Carriage Museum* (Stony Brook, N.Y.: The Museums at Stony Brook, 1987) (no author given).
2. *The Complete Poetical Works of Oliver Wendell Holmes*, ed. H.E.S. (Boston: Houghton, Mifflin, 1895).
3. T. K. Derby and T. I. Williams, *A Short History of Technology* (New York: Oxford University Press, 1960/1975).
4. "A Ride on Macaroni's Coach," *The Museum of Foreign Literature, Science, and Art*, Vol. XI, New Series (Philadelphia: E. Littell & Co., Sept.–Dec. 1840): pp. 182–183. (This article was picked up from the *Spectator* and no author was named).
5. C. Singer, E. J. Holmyard, A. R. Hall, and T. L. Williams, *A History of Technology*, Vol. V (New York: Oxford University Press, 1958), chapter 18.
6. For more on Marcus's automobile, see: S. Kurinsky, "Siegfried Marcus, Part II: The Automobile and the Internal-Combustion Engine," Fact Paper 32-II (New York: Hebrew History Federation, LTD, undated); and J. J. Flink, "Innovation in Automotive Technology," *American Scientist*, 73 (March/April 1985): pp. 151–161.
7. A. Girdler, "First Fired, First Forgotten," *Cycle World* (February 1998): pp. 62–70.
8. *The New Wonder Book of Knowledge*, comp. and ed. Henry Chase Hill (Philadelphia: The John C. Winston Co., 1923).
9. For Glen Curtiss's story, see A. Gridler, "Very First Vee," *Cycle World* (April 2002): pp. 86–89; K. Cameron, "Creative Power: Glenn Curtiss: Inventor, Manufacturer, Racer, Pilot," *Cycle World* (April 2002): pp. 90–92; and L. R. Eltscher and E. R. Young, *Curtiss-Wright: Greatness and Decline* (New York: Twayne Publishers, 1998), especially chapter 2.
10. K. Borg, "The 'Chauffeur Problem' in the Early Auto Era: Structuration Theory and the Users of Technology," *Technology and Culture* 40, No. 4 (October 1999): pp. 797–832.
11. R. Estep, "The Motor Car and Its Owner," *The American Review of Reviews* (March 1909): pp. 336–338.
12. J. M. Fenster, "The Longest Race," *American Heritage* (November 1996): pp. 64–77.
13. W. Greenleaf, *Monopoly on Wheels* (Detroit: The Wayne State University Press, 1961).
14. G. W. Hobbs, B. G. Elliott, and E. L. Consoliver, *The Gasoline Automobile* (New York: McGraw-Hill Book Company, Inc., 1920).
15. G. Stein, *The Autobiography of Alice B. Toklas* (New York: Harcourt, Brace and Company, Inc., 1933), chapter VI.
16. P. W. Laird, "'The Car without a Single Weakness': Early Automobile Advertising," *Technology and Culture* 37, No. 4 (October 1996): pp. 796–812.

17. V. Scharff, *Taking the Wheel: Women and the Coming of the Motor Age* (New York: The Free Press, 1991).
18. M. Lamm, "Model Marketing," *Audacity* 5, No. 1 (fall 1996): pp. 32–37.
19. J. N. Phillips, *Running with Bonnie and Clyde: The Ten Fast Years of Ralph Fults* (Norman: University of Oklahoma Press, 1996), p. 194.
20. E. R. Milner, *The Lives and Times of Bonnie and Clyde* (Carbondale: Southern Illinois University Press, 1996), p. 101.
21. M. Gianturco, "The Infinite Straightaway," *Invention & Technology* (fall 1992): pp. 34–41.
22. B. Yates, "Duesenberg," *American Heritage* (July–August 1994): pp. 88–99.

Chapter 9

1. F. Rowsome, Jr., *The Verse by the Side of the Road* (New York: The Stephen Greene Press/Pelham Books, 1965, 1990).
2. Among the many books on the mimetic, or vernacular, architecture movement are: J. Heimann and R. Georges, *California Crazy: Roadside Vernacular Architecture* (Tokyo: Dai Nippon, 1985); J. Margolies, *The End of the Road* (New York: Viking Press, 1977); C. H. Liebs, *Main Street to Miracle Mile: American Roadside Architecture* (Boston: Little, Brown and Co., 1985); and R. Venturi, D. S. Brown, and S. Izenour, *Learning from Las Vegas* (Cambridge, Mass.: MIT Press, 1972).
3. P. Hirshorn and S. Izenour, *White Towers* (Cambridge, Mass.: MIT Press, 1979).
4. E. Schlosser, *Fast Food Nation* (Boston: Houghton Mifflin, 2001).
5. D. Grassi, R. Bossaglia, and G. Fisogni, *Gasoline* (Milan: Electa, 1995).
6. J. A. Jakle and K. A. Sculle, *The Gas Station in America* (Baltimore: The Johns Hopkins University Press, 1944).
7. T. P. Helms and C. Flohe, *Roadside Memories: A Collection of Vintage Gas Station Photographs* (Atglen, Pa.: Schiffer Publishing Ltd., 1997).
8. H. Leavitt, *Super Highway—Super Hoax* (New York: Doubleday & Company, Inc., 1970).
9. *Books of the Century*, ed. E. Diefendorf (New York: Oxford University Press, 1996).
10. W. Hardy, *A Guide to Art Nouveau Style* (Secaucus, N.J.: Chartwell Books, Inc., 1986), p. 8.
11. J. H. Lienhard, "Biography of L. Prandtl," *Dictionary of Scientific Biography*, Vol. XI, ed. C. C. Gillespie (New York: Charles Scribner and Sons, 1975).
12. For more on Loewy, see: P. Jodard, *Raymond Loewy* (London: Trefoil Publications, Ltd., 1992); R. Loewy: *Pioneer of American Industrial Design*, ed. Angela Schönberger (Munich: Prestel-Verlag, 1990); R. Loewy, *Un Pionnier du Design Américain* (Paris: Centre Georges Pompidou, 1990); or Loewy's autobiography, R. Loewy, *Never Leave Well Enough Alone* (New York: Simon and Schuster, 1951).

Chapter 10

1. O. Wright and W. Wright, "The Wright Brothers' Aëroplane," *The Century Magazine* (Sept. 1908).
2. I talk about many of these early levitations in *The Engines of Our Ingenuity: An Engineer Looks at Technology and Culture* (New York: Oxford University Press, 2000), chapter 8. For more examples and bibliography, see: http://www.uh.edu/engines/airtransportation.htm .

3. Mackworth-Praed, *Aviation: The Pioneer Years* (Secaucus, N.J.: Chartwell Books, Inc., 1990), chapter 2.

4. L. White, Jr., "Eilmer of Malmesbury, An Eleventh Century Aviator," *Medieval Religion and Technology* (Los Angeles: University of California Press, 1978), chapter 4.

5. Three good sources for the history of ballooning are: T. D. Crouch, *The Eagle Aloft: Two Centuries of the Balloon in America* (Washington D.C.: Smithsonian Institution Press, 1983); D. D. Jackson, *The Aeronauts* (Alexandria, Va.: Time-Life Books, 1981); and L.T.C. Rolt, *The Aeronauts—A History of Ballooning, 1783–1903* (New York: Walker and Company, 1966).

6. G. di Campello Della Spina, "Ballooning by Moonlight: Narrative of a Woman's Trip over the Apennines," *The Century Magazine* (May 1907): pp. 253–259.

7. F. Howard, *Wilbur and Orville: A Biography of the Wright Brothers* (New York: Ballantine Books, 1987), chapter 32.

8. A. Post, "The Man-Bird and His Wings," *Cosmopolitan Magazine* (May 1910).

9. Two good general sources on early women fliers are J. Lomax, *Women of the Air* (New York: Ballantine Books, 1987); and C. M. Oakes, *United States Women in Aviation Through World War I* (Washington, D.C.: Smithsonian Institution Press, 1985).

10. In addition to the sources above, many web sites provide good background on Harriet Quimby. See, for example, http://www.allstar.fiu.edu/aero/quimby.htm and http://www.pbs.org/kcet/chasingthesun/innovators/hquimby.html .

11. W. Menkel, "The Airplane—A Year's Marvelous Progress," *The American Review of Reviews* (August, 1911): pp. 336–338.

12. M. B. Rogers, S. A. Smith, J. D. Scott, and C. Shaw, *We Can Fly: Stories of Katherine Stinson and Other Gutsy Texas Women* (Austin, Tex.: Ellen C. Temple, Publisher, 1983). (This book for young people is a useful source, nevertheless.) See also, D. Wedemeyer, "Katherine Stinson Otero, 86, dies: Pioneer Aviator and Stunt Flier," *New York Times* (July 11, 1977): p. 28.

13. J. H. Cockfield, "All Blood Runs Red," *Legacy: A Supplement to American Heritage* (February/March 1995): pp. 7–15.

14. E. H. Freydberg, entry on Bessie Coleman (1896–1926), *Black Women in America: An Historical Encyclopedia*, ed. Darlene Clark Hine (New York: Carlson Publishing Inc., 1993), pp. 262–263. See also, D. L. Rich, *Queen Bess: Daredevil Aviator* (Washington, D.C.: Smithsonian Institution Press, 1993).

15. V. Hardesty and D. Pisano, *Black Wings* (Washington, D.C.: National Air and Space Museum, Smithsonian Institution, 1984). See especially page 6.

16. W. J. Powell, *Black Aviator* (with an introduction by Von Hardesty) (Washington, D.C.: Smithsonian Institution Press, 1994).

17. R. Welch, *Encyclopedia of Women in Aviation and Space* (Santa Barbara, Calif.: ABC-CLIO, Inc., 1998). See also, V. Moolman, *Women Aloft* (Alexandria, Va.: Time-Life Books, 1981).

18. V. Moolman, pp. 43–45.

19. Beryl Markham's story is told by L. S. Lovell, *Straight on Till Morning* (New York: St. Martin's Press, 1987); D. Boyles, *African Lives: White Lies, Tropical Truth, Darkest Gossip, and Rumblings of Rumor—from Chinese Gordon to Beryl Markham, and Beyond* (New York: Ballantine Books, 1989), chapter 3; and by Markham herself in B. Markham, *West with the Night* (San Francisco: North Point Press, 1983; 1st ed., 1942).

20. B. Markham.

21. There is a huge body of literature on Amelia Earhart. Some of these sources include M. S. Lovell, *The Sound of Wings: The Life of Amelia Earhart* (New York: St.

Martin's Press, 1989); V. V. Loomis and J. L. Ethell, *Amelia Earhart: The Final Story* (New York: Random House, 1985); S. Ware, *Still Missing: Amelia Earhart and the Search for Modern Feminism* (New York: W.W. Norton and Company, 1993); and A. Earhart, *Last Flight* (New York: Harcourt, Brace, & Co., 1937). (George Putnam built this latter account of Earhart's last flight from her telegraphed dispatches and published it a scant five months after her disappearance.)

22. A. Earhart, *The Fun of It* (New York: Brewer, Warren & Putnam, 1932).
23. J. Mendelsohn, *I Was Amelia Earhart* (New York: Alfred A. Knopf, 1996).
24. D. McCullough, *Brave Companions: Portraits in History* (New York: Simon & Schuster, 1992), chapter 9.
25. A. Saint-Exupéry, *The Little Prince* (New York, Harcourt, Brace, 1943).
26. A. Saint-Exupéry, *Night Flight* (New York: The Century Co., 1932).

Chapter 11

1. R. Reinhardt, "Day of the Daredevil," *American Heritage of Invention & Technology* (fall 1995): pp. 10–21.
2. W. Menkel, "The Airplane—A Year's Marvelous Progress," *The American Review of Reviews* (August 1911): pp. 336–338.
3. W. J. Boyne. *deHavilland DH-4: From Flaming Coffin to Living Legend* (Washington, D.C.: Smithsonian Institution Press, 1994).
4. R. S. Holland, *Historic Airships* (Philadelphia: Macrae-Smith Company, 1928), chapter XVI, "Around the World."
5. Le Corbusier, *Aircraft* (New York: Universe Books, 1988, reprint of a 1935 English edition).
6. For more on Le Corbusier and his ideas, see F. Choay, *Le Corbusier* (New York: George Braziller, Inc., 1960); and G. H. Baker, *Le Corbusier: An Analysis of Form* (Hong Kong: Van Nostrand Reinhold Co. Ltd., 1984).
7. "Curtiss-Wright Airports: A Nation-Wide Chain of Strategically Located Ports" (New York: Curtiss-Wright Airports Corporation, 1929).
8. A. K. Lobeck, *Airways of America: The United Air Lines* (New York: The Geographical Press, Columbia University, 1933).
9. E. Angelucci, *World Encyclopedia of Civil Aircraft* (English Language Edition) (New York: Crown Publishers, Inc., 1982), See especially chapter 5.
10. S. B. Nelson, "Airports Across the Ocean," *Invention and Technology* 17, No. 1 (summer 2001): pp. 32–37.
11. M. Economides and R. Oligney, *The Color of Oil: The History, the Money and the Politics of the World's Biggest Business* (Katy, Tex.: Round Oak Pub. Co., 2000), p. 151.
12. *Aviation Week and Space Technology* 135, No. 6 (August 12, 1991). See especially O. Wright, "The Future of Civil Flying," p. 61, and A. Earhart, "Putting Air Travel into Mass Production," pp. 108–111, as well as other reprinted material from older issues.
13. R. McG. Thomas, Jr., "Douglas Corrigan, 88, Dies; Wrong-Way Trip Was the Right Way to Celebrity as an Aviator," *New York Times*, December 14, 1995, p. C19.
14. L. S. Reich, "From the Spirit of St. Louis to the SST: Charles Lindbergh, Technology, and Environment," *Technology and Culture* 36, No. 2 (April 1995): pp. 351–393.
15. J. Nance, *The Gentle Tasaday: A Stone Age People in the Philippine Rain Forest* (foreword by Charles A. Lindbergh) (New York: Harcourt Brace Jovanovich, 1975).

16. C. A. Lindbergh, *Of Flight and Life* (New York: Charles Scribner's Sons, 1948), pp. 10–14.
17. See note 16.
18. A. M. Lindbergh, *Gift from the Sea* (New York: Pantheon Press, 1955).
19. See note 5.
20. C. A. Lindbergh, *Autobiography of Values* (New York: Harcourt Brace Jovanovich, 1976), pp. 366–367.

Chapter 12

1. I have done many papers in this area with colleagues and students. See, e.g.: N. Shamsundar and J. H. Lienhard, "Equations of State and Spinodal Lines—A Review," *Nuclear Engineering and Design* 141 (1993): pp. 269–287.
2. J. H. Lienhard and J. M. Stephenson, "Temperature and Scale Effects upon Cavitation and Flashing in Free and Submerged Jets," *Journal of Basic Engineering* 88, No. 2 (1966): pp. 525–532.
3. The following three items deal with protecting against thermohydraulic explosions: Md. Alamgir, C. Y. Kan, and J. H. Lienhard, "An Experimental Study of the Rapid Depressurization of Hot Water," *Journal of Heat Transfer* 102, No. 2 (1980): pp. 433–438; Md. Alamgir and J. H. Lienhard, "Correlation of Pressure Undershoot During Hot Water Depressurization," *Journal of Heat Transfer* 103, No 1 (1981): pp. 52–55; and J. H. Lienhard and G. S. Borkar, Quick Opening Pressure Release Device and Method, U.S. Patent no. 4,154,361, May 15, 1979.
4. *The Boy Mechanic:* Book I: *700 Things for Boys to Do* (Chicago: Popular Mechanics Co., Publishers, 1913).
5. M. Orvell, "A Hieroglyphic World: The Furnishing of Identity in Victorian Culture," *The Real Thing: Imitation and Authenticity in American Culture* (Chapel Hill: The University of North Carolina Press, 1989), chapter 2.
6. C. L. Boone, *The Library of Work and Play: Guide and Index* (Garden City, N.Y.: Doubleday, Page & Company, 1912).
7. A. F. Collins, *The Boy Scientist* (Boston: Lothrop, Lee & Shepard Co., 1925).
8. C. Dickens, *Bleak House* (illustrated by H. K. Browne) (London: Bradbury and Evans, 1853), chapter XIX.
9. A. E. Housman, *A Shropshire Lad* (London: Kegan Paul, Trench, Trübner, and Co. Ltd., 1896), poem XXXVI.
10. G. A. Henty, *Jack Archer: A Tale of the Crimea* (New York: Hurst and Company, 1900).
11. D. C. Beard, *The American Boy's Handy Book* (foreword by Noel Perrin) (Boston: David R. Godine, Publisher, Inc., 1983).
12. H. H. Hickam, Jr., *October Sky* (aka *Rocket Boys*) (New York: Island Books, 1999).
13. J. B. Bullard, "Two Cadiz High School Students Plan to Fire Multi-Stage Rocket," *Martin's Ferry Paper*, ca. 1958.

Chapter 13

1. J. Hope, "A Better Mousetrap," *American Heritage* (October 1996): pp. 90–97.
2. *The Boy Mechanic:* Book I: *700 Things for Boys to Do* (Chicago: Popular Mechanics Co., Publishers, 1913).

3. R. Dale and J. Gray, *Edwardian Inventions: 1901–1905* (London: W. H. Allen & Co. Ltd., 1979).

4. Petroski tells the story of the paperclip in an article: H. Petroski, "The Evolution of Artifacts," *American Scientist* 80 (September/October 1992): pp. 416–420, and also in his book, H. Petroski, *The Evolution of Useful Things* (New York: Alfred A. Knopf, 1992), chapter 4.

5. G. Wise, "Ring Master," *American Heritage of Invention & Technology* 7, No.1 (spring/summer 1991): pp. 58–63.

6. T. Lewis, *Empire of the Air: The Men Who Made Radio* (New York: HarperCollins Publishers, 1991).

7. "Armstrong's Super-Regenerative Receiver," *Scientific American* 127 (September 1922): p. 160.

8. R. C. Hayden, *Black American Inventors* (Reading, Mass.: Addison Wesley Publishing Company, 1972).

9. C. Hallowell, "Charles Lindbergh's Artificial Heart," *American Heritage of Invention & Technology* 1, No. 2 (1985).

10. D. J. Golembeski, "Struggling to Become an Inventor," *American Heritage of Invention & Technology* (winter 1989): pp. 8–15.

11. E. Pace, "Inventor of Instant Camera, Edwin H. Land, dies at 81," *New York Times*, March 2, 1991.

Chapter 14

1. J. L. Heilbron, *H. G. J. Moseley: The Life and Letters of an English Physicist, 1887–1915* (Berkeley: University of California Press, 1974). J. L. Heilbron, "Moseley, Henry Gwyn Jeffreys," *Dictionary of Scientific Biography*, ed. C. C. Gilespie (New York: Charles Scribner's Sons, 1970–1980).

2. H. S. Maxim, *My Life* (London: Methuen & Co., Ltd., 1915), and I. Hogg, *The Weapons that Changed the World* (New York: Arbor House, 1986).

3. H. S. Maxim, "A New Flying-Machine: Maxim's Experiments in Aerial Navigation," *The Century Magazine* XLIV, No. 3 (January 1895): pp. 444–456.

4. H. S. Maxim, "Aerial Navigation: The Power Required," *The Century Magazine* XLII, No. 6 (October, 1891): pp. 829–836.

5. A. P. van Gelder and H. Schlatter, *History of the Explosives Industry in America* (New York: Arno Press, 1972), pp. 778–780.

6. M. Evlanoff and M. Flour, *Alfred Nobel: The Loneliest Millionaire* (Los Angeles: The Ward Ritchie Press, 1969); K. Fant, *Alfred Nobel: A Biography* (New York: Arcade Publishing, 1991, 1993).

7. M. Evlanoff and M. Flour, p. 84.

8. M. Evlanoff and M. Flour, p. 160.

9. W. B. Robinson, *American Forts: Architectural Form and Function* (Urbana: University of Illinois Press for the Amon Carter Museum of Western Art, Fort Worth, Tex., 1977).

10. D. D. Porter, "The Opening of the Lower Mississippi," *The Century Magazine* XXIX, No. 6 (April 1885): pp. 921–952.

11. B. W. Tuchman, *The Guns of August* (New York: Macmillan, 1962).

12. For the fascinating tale of the French use of balloons in 1870, see B. Mackworth-Praed, *Aviation: The Pioneer Years* (Secaucus, N.J.: Chartwell Books, Inc., 1990), chapter 2; and J. Glaisher, C. Flammarion, W. de Fonvielle, and G. Tissandier, *Travels*

in the Air, 2nd and rev. ed. (London: Richard Bentley & Son, 1871), especially the preface.

13. C. Dienstbach and T. R. MacMechen, "The Aerial Battleship," *The McClure Magazine* (August 1909): pp. 422–434.
14. W. Manchester, *The Arms of Krupp: 1587–1968* (Boston: Little, Brown and Company, 1964).
15. W. Cross, *Zeppelins of World War I* (New York: Paragon House, 1991).
16. The story of World War I airplane design is told in many places. I suggest two particularly accessible sources: T. C. Treadwell and C. A. Wood, *The First Air War: A Pictorial History, 1914–1919* (New York: Barnes & Noble Books, 1996); and M. Sharpe, *Biplanes, Triplanes and Seaplanes* (New York: Barnes & Noble Books, 2000).
17. D. Brannon (with Don Greer, Joe Sewell, and Randle Toepfer), *Fokker Eindecker in Action* (Aircraft Number 158) (Carrollton, Tex.: Squadron/Signal Pubs., Inc., 1996).
18. P. Cooksley (with Don Greer and Ernesto Cumpian), *Nieuport Fighters in Action* (Aircraft Number 167) (Carrollton, Tex.: Squadron/Signal Pubs., Inc., 1997).
19. O. H. Peets, "The Fields of France," *The Century Magazine* LXXV (November 1918): pp. 57–53.
20. M. J. Neufeld, "Weimar Culture and Futuristic Technology: The Rocketry and Spaceflight Fad in Germany, 1923–1933," *Technology and Culture* 31, No. 4 (Oct. 1990): pp. 725–752.
21. J. Lomax, *Women of the Air* (New York: Ivy Books, 1987), chapter 13.
22. A. Roca, "Killer Air Ray," *American Heritage of Invention & Technology* 11, No. 1 (summer 1995): p. 64.

Chapter 15

1. Dr. Balraj Sehgal, at this writing, runs the nuclear laboratory at the Swedish Royal Institute of Technology.
2. H. A. Johnson, V. E. Schrock, F. B. Selph, J. H. Lienhard, and Z. R. Rosztoczy, "Transient Pool Boiling of Water at Atmospheric Pressure," *International Developments in Heat Transfer* (New York: ASME, 1963), pp. 244–254.
3. For background on the role of Hahn and Meitner, see R. Rhodes, *The Making of the Atomic Bomb* (New York: Touchstone Books, 1986); or E. A. Vare and G. Ptacek, *Mothers of Invention* (New York: Quill, 1987).
4. S. L. Del Sesto, "Wasn't the Future of Nuclear Energy Wonderful?" *Imagining Tomorrow: History, Technology, and the American Future*, ed. Joseph Corn (Cambridge, Mass.: MIT Press, 1986), chapter 3.
5. T. C. Reeves, *The Life and Times of Joe McCarthy* (New York: Stein and Day, 1982).
6. J. Kerouac, *On the Road* (New York: Penquin Books, 1955).
7. M. Theado, *Understanding Jack Kerouac* (Columbia: University of South Carolina Press, 2000).
8. A. Ginsberg, *Collected Poems: 1947–1980* (New York: Harper & Row, Publishers, 1984), p. 131.
9. A. Ginsberg, p. 154.
10. This story was told to me by Professor Dorothy Z. Baker, University of Houston English department.
11. J. Stoll and E. S. Connell, Jr., *I Am a Lover* (Sausalito, Calif.: Angel Island Publications, Inc., 1961).
12. Two good sources on the history of the digital computer are T. R. Reid, *The Chip: How Two Americans Invented the Microchip and Launched a Revolution* (New York:

Simon and Schuster, 1984); and M. S. Malone, *The Microprocessor: A Biography* (New York: Springer Verlag, ELOS, 1995).

13. K. Clark, *Civilisation: A Personal View* (New York: Harper & Row, Publishers, 1969), p. 347.

Chapter 16

1. Z. G. Pascal, *Endless Frontier: Vannevar Bush, Engineer of the American Century* (New York: Free Press, 1997).
2. Z. G. Pascal, p. 380.
3. V. Bush, "As We May Think," *The Atlantic Monthly* 176, No. 1 (July 1945): pp. 101–108.
4. L. Owens, "Vannevar Bush and the Differential Analyzer: The Text and Context of an Early Computer," *Technology and Culture* 27, No. 1 (January 1986): pp. 63–95.
5. See note 3.
6. See note 3.
7. E. L. Eisenstein, *The Printing Press as an Agent of Change*, Vols. I and II (Cambridge: Cambridge University Press, 1979). See especially part I, chapter 1, "The Unacknowledged Revolution."
8. *Dürer: Des Meisters Gemälde, Kupferstiche, und Holzschnitte in 447 Abbildungen* (Mit einer Biographischen Einleitung von Dr. Valentin Scherer) Klassiker der Kunst in Gesamtausgaben, Vierter Band (Stuttgart und Leipzig: Deutsche Verlags-Anstalt, 1904), p. 131.
9. H. D. Thoreau, *Journal: Vol. 5: 1852–1853*, ed. Patrick F. O'Connell and Robert Sattelmeyer (Princeton, N.J.: Princeton University Press, 1997). See the entry for July 16, 1852.
10. A. C. Clark, *The Fountains of Paradise* (New York: Harcourt, Brace, Jovanovich, 1979).
11. C. Sheffield, *The Web Between the Worlds* (New York: Ballantine Books, 1979).

Index

Index

Index

Index

Ford, Henry, 104, 122, 125, 126–27, 131–33, 144
Ford, John, 70
Ford automobiles, 124, 130, *130*
Ford Company, 187
Ford Trimotor airplanes, 182, *182*
Ford V-8, 132–33
forts, 227, *228*, 229
The Fountains of Paradise (Clarke), 266
4'33" (Cage), 254
framing techniques in construction, 87–89
Frankenstein: Or the Modern Prometheus
    (Shelley), 2
freedom of movement, 116–17, 122
French Lick Springs spa, 35
Fulton, Robert, 3–4, 120

Gable, Clark, 135
Galapagos Islands, 13–14
Galileo, 62
Gann, Ernie, 171
Garnerin, André, 155–56
Garroway, Dave, 164
gas stations, 144–45
gasoline, 142–45
Gates, Bill, 131
Gauchot, Paul, 153
Gauguin, Paul, 64
Geanangel, Alan (Russ), 203
gear shifting in automobiles, 128, *129*
General Motors, 130
genetic engineering, 263
genius, **53–63**, 198
Geyser, Albert, 45
Gianturco, Michael, 134
Gibbs, J. Willard, 125
Giffard, Henri, 153–54
Gift from the Sea (Lindbergh), 188
Gillette, King Camp, 101–5, *103*, *104*, 111
Ginsberg, Allen, 252–53, 254
Girdler, Allan, 121
glass heart, 214
gliders and gliding. *See also* airplanes; flight
    in *The Boy Mechanic,* 192–93, *193*
    Hanna Reitsch, 241–42
    in *Popular Mechanics,* 199
    precedents for, 152
    of Wright brothers, *152*, 207
Goddard, Robert Hutchings, *240*
Goethe, Johann Wolfgang von, 57–58, 60
Gogh, Vincent van, 64, 77–78
Gold King Company, 14
The Gossips (Remington), *83*
Gothic themes in architecture, 109
Gray, Joan, 207
Greenleve, Block & Co., Galveston, Texas, *28*
Guernica (Picasso), 78–79
Guest, Amy, 168
A Guide to Art Nouveau Style (Hardy), 147
The Guns of August (Tuchman), 229
Gutenberg's press, 263
gyrocompasses for airplanes, 61

Hahn, Otto, 246, 247
hair removal by X-ray, 45–46
Hallidie, Andrew, 33–34
Hallowell, Christopher, 214
Haloid company, 215
Hardy, William, 147
Harley, Bill, 122
Harper's Monthly, 70

Haviland, John, 28
Hayden, Sophia, 18
hearts, artificial, 214–15
Heisenberg's Uncertainty Principle, 42, 50
Hemingway, Ernest Miller, 251
Henderson, Brooks, 144
Henry, Joseph, 12
Henty, George A., 200, *201*
The Herd at Night (Remington), 71
Herzig, Rebecca, 45
Hickam, Homer, 201–3
hieroglyphic world, 195
Historic Landmarks Foundation, 36
Hitler, Adolf, 241
HMS *Dreadnought,* 29–30, *30*
hobble skirts, 158
Holmes, Oliver Wendell, 118
Home Insurance Building, 92, 93
"homemade" designation, 196
Hop Harrigan radio serial, 185
Hope, Jack, 204
Hopper, Edward, 78
House Un-American Activities Committee
    (HUAC), 253
Housman, A. E., 199, 201
Houston, Texas, 112, *113*
Houston Astrodome, 35
Howard, Fred, 158
Howe, Elias, 7
Howl (Ginsberg), 252
Huckleberry Finn (Twain), 5
Hughes, Thomas, 58
The Human Drift (Gillette), 101, 102, *103*
The Human Use of Human Beings (Wiener), 56
Hupmobile automobile, 150
hydraulic elevators, 91, *91*, 96
Hymn to St. Cecilia (Britten), 248

I Am a Lover (Stoll), 253
Ibn Firnas, 154
Icarus, 154
Illinois Tower, 106
Imaginary Landscape no. 4 (Cage), 254
impressionism, 64, 70
incandescent lighting systems, 59
Indian Motorcycle Company, 121, 122
individualism, 196–97
industrial arts, 195
Indy 500, 134
information flood, 260–61
infrastructure, 24, 26, 36
initial conditions, dependence on, 21–22
INTEL, 256
interchangeable parts, 7
Internet, 261
Interstate Highway System, 125, 145–46
invention and inventors, **204–18**
iron, 27–29, *28*
ironclads, 10

J-model Dusenbergs, 136
Jack Archer: A Tale of the Crimea (Henty), *200*
James, William, 13, 14
Jeanneret, Charles-Édouard (Le Corbusier),
    179–80
Jencks, Charles, 101, 103
John A. Roebling and Sons, 183
Johnston, Frances Benjamin, 74–75
Jordan automobiles, 130, *131*
Joyce, James, 52

Index

Index

music, 51–52, 254. *See also specific compositions and composers*
musket-ball mold and ball, *190,* 190–91
My Life (Maxim), 220, *221, 222*
mystery and technical revolution, 48–49

NASA, 266
Nast, Thomas, 26, *26*
National Academy of Design, 76
negative feedback, 55
Nelson, Steward, 183
New Orleans airplanes, 178
New York City, 99
Nickerson, William, 102
Niépce, Joseph, 73
Nieuport 28 airplanes, 235
nitroglycerin, 225
Nobel, Alfred, 225–27
Nobel prizes, 226–27
North American Hiroshima maiden syndrome, 46
Noyce, Robert, 255–56
nuclear-fusion energy, 246–47
nuclear power plants, 192
Nude Descending a Staircase (Duchamp), 51, 78, *78*

O-rings, 209–10, *210*
Oberth, Hermann, 240–41
October Sky (film), 201
Odell family, 138
oil platforms, 184
Oldfield, Barney, 134, 173
Olds, Ransom Eli, 126, 127
Oldsmobiles, 124, 128, 129
Olmsted, Frederick Law, 17
omnibuses, 107
On the Road (Kerouac), 251–52
One Minute of Silence (Batt), 254
O'Neil, John, 203
ophthalmoscopes, 47
Oppenheimer, J. Robert, 248, 259–60
Orvell, Miles, 195
Otis, Elisha, 28, 90–100, *92*
Otis Elevator Company, 98

Packard automobiles, 130
Painter, William, 101
paperclips, 208–9
parachutes, 155–56
Paris Exhibition (1889), 17, 94
patents, 204, *206,* 207–8
Patterson, Lucy, 203
Peace Prize of the Nobel prizes, 226–27
Pearl Street Power Station, New York, 59
Penaud, Alphonse, *20,* 153–54
The People's Corporation (Gillette), 104
perfusion pump, *214*
perpetual-motion machines, 207
Perrin, Noel, 201
Peshtigo, Wisconsin, 84–85
Petroski, Henry, 208
Phaedra (Plato), 155
phaetons, 118
Philadelphia Fair, 17
phonographs, 59
phosphorescent light, photos from, *61*
The Photo-Secession, 74–75
photography, 73–75, 193, 217

Picasso, Pablo, *76,* 78–79
pipe bombs, 193–94
Planaphore model airplane, 153, *153*
Planck, Max, 39–40, 41, 43, 49, 246
platform construction, 88
Plato, 245
Playa del Rey speedway, 134
poison gas, 242
polarization of light, 216–17
Polaroid, 216–17
camera, 217
Pony Express, 23
Popular Mechanics, 262
populist movements, 122, 132
postimpressionism, 64, 73
postmodern era, 1, 150
Powell, William J., 165
Prandtl, Ludwig, 149, *149*
Premier automobiles, 128
press of Gutenberg, 263
Prince, Jack, 134
protactinium, 246
psychology of robots, 55
Pure, 144
Putnam, George, 168–69

quantum mechanics, 42–43
Quimby, Harriet, *159,* 159–60, 167, 170

racing cars, 124–25, *134,* 149, 175, *175*
racism, 8–9, 213
radio, 18, 60, 144, 199, 211–12, 262
radio diodes, 207
radiology, 16–17
Raiche, Bessica, 159
rail systems, electric, 58
railways, 3, 7, 15, 23, 117, 146
razors, 102
Red Baron (Manfred von Richthofen), 234
red scare, 248–51, 259
refrigeration, 61
Reich, Leonard, 186–89
Reiser, Stanley, 46, 47
Reitsch, Hanna, 241–42, 243
Remington, Frederic
career, 65–66, 69–71, *71,* 82, 83
paintings, *66, 67, 68, 83*
Reno, Jesse, 97–98, *98*
Rhodes-Moorhouse, 233–34
Richthofen, Manfred von (the Red Baron), 234
Rickenbacker, Eddie, 136
rifles, 7
Rillieux, Norbert, 8
riverboats, 4, 8–9, 9–10
roads and highways, 117, 125, 133, 136, **137–42**
Roadside Memories: A Collection of Vintage Gas Station Photographs (Helms and Flohe), 145
robot psychology, 55
Roca, Alexander, 242
Rockefeller Center, 109
Rockefeller Differential Analyzer, 260
Rocket Boys (Hickam), 201–3
rocketry, *240,* 240–42, 255, 258, 265–66
Rodriguez y Robertson, Garcia, 30
Roebling, John, 31–32, 33
Roebling, Washington, 33
Roentgen, Wilhelm Conrad, 16, *43,* 43–46, 227
Roethke, Theodore, 252

Index

Index